THE SEASIDE

Also by Madeleine Bunting

NON-FICTION

Labours of Love: The Crisis of Care

Love of Country: A Hebridean Journey

The Plot: A Biography of an English Acre

Willing Slaves: How the Overwork Culture
is Ruling Our Lives

The Model Occupation: The Channel Islands under
German Rule 1940–45

FICTION

Ceremony of Innocence

Island Song

THE SEASIDE

England's Love Affair

MADELEINE BUNTING

GRANTA

Granta Publications, 12 Addison Avenue, London W11 4QR

First published in Great Britain by Granta Books, 2023

A CIP catalogue record for this book
is available from the British Library.

1 3 5 7 9 10 8 6 4 2

ISBN 978 1 78378 717 3
eISBN 978 1 78378 718 0

Typeset by M Rules

Printed and bound by CPI Group (UK) Ltd, Croydon, CR0 4YY

www.granta.com

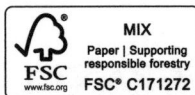

*Dedicated to those I met on my journeys
whose love of place was inspiring and who
committed time and energy to ensuring
a better future for seaside resorts.*

*Also dedicated with great affection to Kate Sebag
in gratitude for a friendship now approaching four
decades. Along the way, we've explored
a few coastlines.*

CONTENTS

SOME OF
ENGLAND'S
SEASIDE
RESORTS

NEWCA

Morecambe
Blackpool
Lytham St Annes
Southport
Crosby
New Brighton

PRE
MANCH
LIVER

BIRMIN

SWANSEA
CARDIFF
BR

Clevedon
Weston-super-Mare
Watchet
Minehead
Ilfracombe
Woolacombe
Braunton

Padstow

Paignton
Torquay
Weymouth
Bourn

Whitley Bay
South Shields
Saltburn-by-the-Sea
Runswick Bay
Whitby
Scarborough
Filey
Bridlington
Hornsea
Withernsea
Cleethorpes
Mablethorpe
Sutton on Sea
Chapel St Leonards
Ingoldmells
Skegness

Hunstanton
Cromer

INGHAM

Great Yarmouth
Lowestoft
Southwold

Butlins

Felixstowe
Harwich
Frinton-on-Sea
Clacton-on-Sea
Jaywick
Southend-on-Sea
Canvey Island
Margate
Broadstairs
Ramsgate
Dover
Folkestone

LONDON

ROCK
BO-NUT
FLOSS
OPEN

HAMPTON

Hastings
Bexhill-on-Sea
Eastbourne
Brighton and Hove
Worthing
Bognor Regis

Illustrations

THE SEASIDE

Prologue

I was in Felixstowe for a book festival and, before my session, I slipped away for a quick dip in the sea; it was a hot sunny day and the beach was busy with families, though not many fancied the chilly, tea-coloured water. I gave my talk, still sticky with salt. Afterwards, talking to the organisers, my interest was piqued by a detail of local history. Mrs Wallis Simpson had retreated to one of Felixstowe's glamorous hotels in the midst of the abdication crisis of 1936. I had passed the huge red-brick hotel on my way to the beach; now converted to flats, its lavish splendour was still recognisable in the sweep of its drive and the grand stone portico. On the train back to London, I thought of Simpson in her hotel room, perhaps drinking a cocktail, staring at the North Sea and listening to the waves break, as her future husband's reign unravelled and the country was shaken by constitutional crisis.

The anecdote lingered in my mind in ways I found hard to fully explain: partly, because I sensed that history was no longer made in sleepy Felixstowe. Its pretty terraced streets brimming with flowers offer a peaceful retirement, while the large grand houses once used by wealthy London families are flats or care homes. Perhaps their previous inhabitants are now waterskiing off Greek islands for their summer holidays; those who have retained their affection for Suffolk head further up the coast to Aldeburgh's marshes, quaint pubs and Southwold's retro charms. They don't want the docks of a major commercial port on the horizon as they sit in a deckchair and look out to sea. John Betjeman captures this mournful sense of endings in 'Felixstowe, or The Last of Her Order', in which a nun reflects on the demise of her convent.

> *In winter when the sea winds chill and shriller*
> *Than those of summer, all their cold unload*
> *Full on the gimcrack attic of the villa*
> *Where I am lodging off the Orwell Road.*

Felixstowe, at the end of what is effectively a peninsula created by two estuaries, felt adrift – perhaps happily so? – from modern Britain, its population disproportionately elderly and white, and predicted to become even more aged in the coming decades. Like other English seaside resorts, Felixstowe is no longer described as glamorous or fashionable, and the social cachet of Simpson's day lies in the past. It is haunted by its heyday; few places reverberate so noisily with the ghosts of their history as the English seaside resort.

Shortly after, a friend recounted how she had been walking in North Yorkshire and her route had taken her through Scarborough, another town once known for its sophistication and elegance. We had both known the place as children, but what she described was not the magical place of my memories but a town struggling for a viable future. 'Shabby and rundown,' she lamented. It was a jolt. I hadn't visited in decades. My memories were those of a small child: the stretches of golden sand and blue sea; the thrill of the donkey's measured plod as I swayed unsteadily, gripping to the reins, the animal's distinctive smell, its ears twitching to deter the flies. Later, the treat was candyfloss, melting 99 ice creams and, on the coach journey home, pink and white rock, my hands sticky with sweetness. The childhood idyll I remembered from the 1960s has changed. And Scarborough is not alone. For over two decades, seaside resorts have been found to have the worst levels of deprivation in the country, while a raft of shocking indicators – from poor health, shortened lives, drug addiction, high debt, low educational achievement to low income – demonstrate the blighted communities which cluster along English coastlines. Why has this accumulation of social and economic challenges proved so persistent? Why was the country's deepening inequality so starkly evident here? And why had so little been done to reverse the decline; was it ineffective policy, lack of government effort or an indifferent public?

I couldn't find an account that adequately answered my questions. This was a problem sitting in plain sight; the decline was well known, yet seemed to prompt little more than a resigned shrug. It was as if an entrenched pattern of poverty came down to the availability (or not) of sunshine, and, given the legendary stoicism of the English in dealing with their damp and unpredictable climate, there was little to be done.

My imagination was captured by these two related but distinct themes. Firstly, how do these resorts live with their noisy, vibrant pasts and their powerful, capacious stock of memory, nostalgia, decline and identity? My childhood Scarborough memories are not unusual; part of being English is an emotional tie to the seaside and a collection of shared experiences: caught by a downpour, sand in the sandwiches, cold seas, sheltering from the wind, and then the opposite, those precious days when the sun is high in a perfect blue sky, the sea is sparkling and the pleasure is all the more intense for being unexpected. Millions flock to every coastline annually. Loved and loathed, the seaside resort rarely provokes indifference: it is England's great and turbulent love affair. Secondly, and of a very different order: what are the harsh drivers of the entrenched deprivation found in these places, and how have they become tropes of national decline? Now alert, my attention was caught by news stories relating to these places – and I couldn't but notice how frequently they cropped up, a barometer by which the nation took stock of its weather, climate crisis and cherished holiday habits. I needed to see these places for myself, and a plan took shape to travel England's edges, visiting resorts to get the measure of how they shape a country and its self-image.

The English invented the seaside resort over three centuries, with such imagination, inventiveness and enthusiasm that it became one of their most successful exports. Seaside resorts achieved an iconic national popularity and became central to England's understanding of itself. In the early eighteenth century, they evolved as popular spas with medicinal sea bathing, and then morphed into places of pleasure, entertainment, fantasy, magic and adventure. England fell in love with its edges. Some of this passion is to do with simple geography, since nowhere in England is more than seventy miles from the sea, and its relatively small landmass has a striking variety of coastlines: cliffs, coves, pebbled shore, wide sandy beaches, salt marshes, inlets, and estuaries cutting deep inland. Some of it is because of the mythology of England as an island – despite the fact that we share an archipelago with two other nations. By the latter part of the nineteenth century, railways carried dozens of packed trains daily from every major city to the seaside, and continued to do so for the next seventy years. By the late twentieth century, motorways were regularly clogged with traffic jams as millions made their pilgrimage to the coast. Wales and Scotland developed resorts such as Tenby and Llandudno, Troon and Musselburgh, but they couldn't match the number, size and scale of the attractions of those in England.

The seafront was an English invention, as was the promenade, the pier and even the very idea of the beach as a place of leisure, now a global commonplace. In England, the seaside became a unique public space, free of state, religious or military purpose; unlike in some parts of Europe, most of the foreshore has always been publicly accessible, with only a relatively small proportion of it privately owned. That access is a quirk of legal history by dint of the Crown Estate's ownership of the foreshore – technically defined as the area between high and low tide. (The legal position is a lot less clear than is commonly believed: in a 2015 Supreme Court case, the judges could not make up their minds whether bathers on a Newhaven beach were in fact trespassing. Despite this legal uncertainty, it is commonly accepted that beaches can be used for recreational purposes.) It is hard now to fully appreciate the significance of free public space at the seaside to the nineteenth-century visitor. England was rapidly urbanising and, as migrants moved from countryside to city and the enclosure of land as private property gathered pace, there were few places to escape the foetid smog of intensely populated cities. Their residents literally gasped for breath, their children were desperate for open space and light. The rapid expansion of Blackpool, Morecambe, Skegness, Southend and Clacton owed a lot to the appetite for fresh air, and made much of it in their advertising. Urban parks were comparable, but the freedom they offered was constrained by opening times, railings, gates and park-keepers. Meanwhile, the English rural idyll became a middle-class privilege (and it's been a long, slow process to claw back some portion through the setting up of National Parks and access for walkers). The one space left open to all was the beach. While the European city had the square, the English had the seaside promenade, pier and beach.

Weymouth pier is the setting for the final moving chapter of Kazuo Ishiguro's novel *The Remains of the Day*, the conviviality of the crowds gathering for the evening lights to be lit is a rare moment of warmth in the book's chilly portrayal of the corrupt English elite and the crushing class system. The seaside was one of England's few democratic spaces; no wonder, then, that the Labour politician David Miliband plucked at this national heartstring with his declaration in 2007 that the last remaining privately owned parts of the coast should be opened up to the public: 'England's coastline is a national treasure ... We want to create an access corridor so that people can walk the entire length of the English coast ... We are an island nation,' he announced, invoking the old myth. 'The coast is our birthright and everyone should be able to enjoy it.'

No European capital can lay claim to as many coastlines as London: from Southend and Clacton in Essex, to Margate and Ramsgate, Deal and Folkestone in Kent, to Hastings, Eastbourne, Brighton and Bognor Regis in Sussex – all an easy day trip by train or car. Not to mention the delights further afield, such as the Isle of Wight and Bournemouth. London has within its reach hundreds of miles of seaside and dozens of resorts for those in need of space, light and fresh air. Similarly, Liverpool and Manchester recruited the coast of North Wales to add to their Lancashire resorts. Even now, in an age of cheap air travel, many of the biggest resorts continue to be among the most popular visitor attractions in the country; before the Covid pandemic, Blackpool had eighteen million visitors a year and Brighton 9.5 million, comfortably beating Buckingham Palace's half a million, or the British Museum's 6.2 million. The love affair with the seaside continues, despite deprivation and economic precarity.

Today, 36 per cent of the UK's population live within five kilometres of the sea, and 63 per cent within fifteen kilometres; in Europe, only Italy claims a comparable population distribution around its coasts. Historians and writers stress the significance of England's intimate relationship with its coast. Given that the landscape lacked grandeur or scale, historian John Gillis suggests that the sea offered the English 'a vision of themselves as an heroic nation in a way that their tenuous terrestrial domains could never do'. He quotes Robert Louis Stevenson: 'if an Englishman wishes such a [patriotic] feeling it must be about the sea . . . the sea is our approach and bulwark; it has been the scene of our greatest triumphs and dangers; and we are used in lyrical strains to claim it as our own'. 'National pride reinforces the emblematic status of the seaside,' writes cultural historian Ursula Kluwick. One characteristic of England's love affair with its seaside is its fond belief that it reflects some of its best qualities. In Travis Elborough's affectionate portrayal, *Wish You Were Here*, he sums these up as egalitarian, tolerant, uncomplaining. Images of seaside bank holidays in the media

Ralph and Jenny Stephenson
Together by the Splashing Up Wall
Sit and look out to the sea,
With a tartan rug and flask of tea,
Feel a hug, share a smile
Be warmed by their love for a while.

regularly portray the stoicism of the English, as holidaymakers brave the weather, put up with the absence of comfort, meagre portions of sun, and demonstrate their appetite for making the best of things. Seafront chat about the weather usually concludes with the cheery comment, 'It could be worse.' At the seaside, in the pouring rain, everyone can believe they are 'in it together' – as politicians were fond of claiming about the Covid pandemic.

In 1925 the literary critic Sir Henry Newbolt made the grandiose claim that: 'An English child ... cannot have long attained the power to read before becoming aware that the sea is his boundary, his safeguard, the only highroad of his food supply and his foreign travel ... sea life is essentially part of national life, part of its daily course, part of its record, part of its imaginative experience.' For the vast majority of English children, that encounter with the sea is at a resort on a crowded promenade or a roller-coaster ride. The American novelist Paul Theroux maintained in 1981 that 'no other country has become so identified with its coasts as this island nation', and embarked on a seaside journey which formed his grim account of decline in *The Kingdom by the Sea*.

Theroux's book depicts the traumatic moment when the era of mass English seaside holidays was coming to an abrupt end. Forty years of decline later, that national identification with the English seaside is losing traction with younger generations, for whom the Costa del Sol, Ibiza and Florida may be as familiar as their nearest English resort. With that fades the quintessentially English memories which have been a common reference point across generations: donkey rides, Punch and Judy, the shortcomings of modest, family-run hotels, Kiss-me-quick hats, lettered rock and candyfloss. For much of the twentieth century, summer holidays followed a remarkably similar format, changing only in small details, until the arrival of package holidays and guaranteed Mediterranean sunshine in the 1970s. The seaside resort may no longer be part of common shared experience in the way it was, but some part of Theroux's point persists; the quantity of flags along numerous coastlines expresses a national assertiveness which feels tense and insecure, and is – appropriately enough – edgy.

I became intrigued by the idea of what gets exposed at the edge, what unravels or frays. Our boundaries, borders and coastline have moved centre stage in the UK's recent turbulent political history, and a sequence of national crises, from Brexit to Covid-19 and the growing flow of migrants arriving on Kent's shores, has seen an intense focus on defining, maintaining and securing edges. The backdrop to my journey includes the high votes in favour of Brexit in almost all seaside resorts – with the exception of Brighton. Many coastal areas, such as Castle Point (Canvey), Tendring (Clacton) and the Isle of Thanet (Margate and Ramsgate), had the highest proportion of Leave votes. While much of the country was divided fairly evenly between Remain and Leave, the argument was clear-cut in resorts such as Blackpool and Great Yarmouth, which had huge majorities in favour of Brexit. Yet this attracted less comment and analysis than the Leave-voting deindustrialised regions inland. My journey was punctuated by reflections on the Leave campaign's narrative of freedom and independence, its anxiety about precarious edges, and an inchoate sense of loss – of a past and of a country.

Searching amongst the tidewrack for insights into these historic shifts in England, I found a fitting metaphor in Swanage in Dorset. The nineteenth-century property developer George Burt brought architectural salvage from London and installed it round his home town: many bollards, lamp posts and watchtowers are still inscribed with the names of London boroughs.

In subsequent decades, artists and critics hailed Swanage's eclectic street furniture as surrealism ahead of its time. Burt's collecting was the corollary of how the tides deposit rubbish and seaweed in the long line of tidewrack at high water. In England things and people emigrate to the edge, through drift, collection or propulsion; the coastline reveals and exposes as well as defines England.

The academic Alex Niven proposes an intriguing argument about the coastal edge. He suggests that a void sits at the heart of England's national identity, and in his book *New Model Island* he argues that during the key period for the development of nationalism (1750–1850) across Europe, England was absorbed in an altogether contrary task, that of effacing itself in order to dominate its neighbour nations of Wales, Scotland and Ireland. While Italians, Germans and Irish developed their sense of national identity, recruiting poets, writers, artists, statesmen, philosophers and generals, England invested its efforts in a multinational Great Britain and its empire. It had no revolutionary movement which developed and defined a popular concept of Englishness, and the result is a vagueness of national identity; confusingly, 'English' was used (and often still is) interchangeably with 'British'. In this absence, pre-industrial English folklore persisted, in which Niven suggests certain themes keep reappearing; one is that 'to be English is to feel hemmed in, straitjacketed, resentful of neighbours' and to have a 'feeling of confinement'. His thesis has particular resonance on the coast, a border which has reassured as a fortification and impeded the flow of ideas and people from the Continent. In the famous speech of John of Gaunt in *Richard II*, William Shakespeare offers a neurotically defensive definition of England.

This fortress built by Nature for herself
Against infection and the hand of war,
This happy breed of men, this little world,
This precious stone set in the silver sea,
Which serves it in the office of a wall,
Or as a moat defensive to a house,
Against the envy of less happier lands.

If English nationalism feels insubstantial and hollow, as Niven argues, then the physicality of the edge – where land meets sea – acquires increased significance as a point of definition, even more so when France is a visible smudged grey line on the horizon. The edge can become a tense place, where old fears of invasion can be easily stirred, as they have been in recent years by irresponsible political leadership when dealing with cross-Channel migrants. England uses its coastline to provide a felt sense of Englishness: an eclectic amalgam of rain, tea, fish and chips, beer and retro beach huts. Repeatedly, the seaside resort is recruited to play a distinctive role, serving as a stage or screen on which national preoccupations, anxieties and insecurities are exhibited; writers, artists and film-makers are drawn to the rich pickings of symbolism found on the seafront and its piers, ballrooms, winter gardens and amusement parks. By the latest tally, Blackpool has been the setting for 250 films, Brighton has clocked up an impressive 563 and Margate 117, and all these figures will be out of date by the time this book has reached publication.

This story of England's coastal edges has also to be set within the context of the climate emergency. Many coastlines are fragile and prone to erosion; cliffs slip and crumble. By definition, a beach is always on the move, as the sand is washed away and replenished. Longer-term trends triggered by the climate crisis bring new threats, such as vanishing beaches, flooding, and accelerated erosion. Increasingly harsh storms pound sea defences, and every winter the images of enormous waves which dwarf the human-made constructions of harbours and seafront promenades both thrill and intimidate. England has a long history of battling with the sea, shoring up its coastline with timber and concrete and engineering flood defences. The prediction is that the cost and effort will only intensify; as this book was going to press, new research assessed that as many as 200,000 homes may have to be abandoned, with whole towns relocated inland because of sea-level rise. Given England's historically mild climate and modest landscape, the sea has been the one place of encounter with the natural world's destructive, uncontrollable power.

Travelling these edges, hopping from one seaside resort to another, I was curious to visit and see for myself places I had only ever heard of. I had a set of questions about them and what they revealed of England, but the journey was also provoked by my affection for the seaside. I love beachcombing, hiding from the wind behind a groyne with a Thermos flask, watching seagulls, eating hot fish and chips, cold-water swimming, and I love the wide horizon, no matter how brown and chilly the sea. It never fails to reach deep into my soul. I decided to begin in Scarborough, the resort closest to where I grew up in North Yorkshire, and the coastline where my own love affair with the sea began.

Between lockdowns provoked by the Covid pandemic, I managed to reach just over forty resorts – about a third of England's resorts, and a few in Wales. (My definition of a resort is a coastal town visited by many to enjoy its beach and attractions, and if my estimates of the total number are vague, it is because there is still discussion amongst policymakers and academics on precise

definitions.) And, yes, I ate a lot of fish and chips – in many places, there was little else available (apart from burgers, and I don't eat meat); if the theme becomes repetitive, so was the experience. I pounced on a lettuce leaf in a Blackpool sandwich shop with relief.

My focus is the big resorts which have dominated England's seaside history, rather than the picturesque charm of popular former fishing villages – these form a story of their own; as a result, Norfolk, Suffolk and much of Cornwall are missing. Some might find the gaps – Great Yarmouth, Bournemouth, Whitley Bay, Newquay – lamentable, and I apologise unreservedly to those who find their favourite, much-loved resort overlooked, but my aim was not to be comprehensive in an almanac style, but to trace an extraordinary national phenomenon of boom and decline, reinvention and struggle. I also wanted to understand how these resorts sit in their landscape – how they emerged from marsh or found themselves perched on clifftops – so each visit took in the shape, history and environmental challenges of the wider coastline. I wandered along some obscure parts of England's crinkled shoreline amongst mud, tidewrack, rubbish and marshes, and yet, even there, I came across people who cherished these neglected parts of our coast; after these forays, I would find my way back to the nearby resort, with its glitter, bright lights, crowds and pungent smell of hot fat.

The coastline is a place of constant flux. It is the meeting point between two elements, earth and water, and is the archetypal liminal space; the Latin word *limen* means threshold, a point of entering and departure, a place of transition. This concept repeatedly emerged in my research. Liminality has a quality of ambiguity and uncertainty, suggests the French ethnologist Arnold Van Gennep, who first coined the word; he goes on to argue that in liminal places social hierarchies, habits and routines can be temporarily set aside. With that comes disorientation and nostalgia, but also the possibility of new perspectives. Another theorist describes how people's response to liminality enlarges their sense of agency and offers a 'sometimes dramatic tying together of thought and experience'. I immediately recognised this point; at times of confusion and intense emotion, my impulse is always to head to the coast, and there I invariably find in the expansive surface of the sea and the rhythm of breaking waves some relief. In interviews, I found many others who also described how the seaside offers clarity, solace and resolution.

The Franciscan priest Richard Rohr advises that liminality is 'where we are betwixt and between the familiar and the completely unknown. There alone is our old world left behind, while we are not yet sure of the new existence. That's a good space where genuine newness can begin. Get there often and stay as long as you can, by whatever means possible.' Resorts have become places of ritual, as the English connect at pivotal points in their lives with their grey and usually muddy seas: ashes are scattered, padlocks attached to pier railings to declare true love, and bunches of dried flowers are ubiquitous on promenade memorial benches. Liminality is evident in the multiple ways resorts have developed: their love of memory, indulgence of appetite and fantasy, and encouragement of adventure and experimentation. What emerges can be carnivalesque, unruly, creative and transgressive, and jostles with constant efforts to maintain order, tidiness and respectability.

Early in my research, as the plans for my journey were taking shape, I made one of my regular forays to the north Kent coast; from where I was then living in east London, it was the nearest place to swim.

The waters of the Thames Estuary lap brown against the brilliant lime green of the seaweed-covered sea wall. It is overcast and cold, and I retreat into my thick coat, shivering slightly in anticipation, because soon I'm going to strip off and my skin will pimple with goosebumps in the stiff westerly wind, and then, however implausible it now seems, I'm going to walk into that sea. I'm going to push off from the uneven pebbles and launch myself into the water, leaving steady ground. I'm going to feel this meeting point between land and water in every part of my body. I rest my eyes on the water and watch how the wind worries its surface, listening to the modest waves breaking and their rattle as they withdraw.

Behind us, the playground and cafe of Reculver are busy, and the ruins of the medieval monastery and the Roman fort are a reminder that this Kent coast has had multiple lives. The empty arches of the old abbey windows look out, sightless, on the estuary, protected from the powerful currents by concrete walls and 'rock armour', the huge granite blocks gouged out of quarries and heaped up along the crumbly low bank. At this point, the land's edge may be ancient, but it is fragile, requiring solicitous reinforcements to see its next century through.

The tide is going out, exposing rough sand. The shore shelves gently, perfect for quick immersion. Unbuttoning coat, pulling off gloves, hat, jumpers, jeans and boots: in the cold speed is essential. Then the feel of icy waves on my legs, lightly buffeting hips and stomach. Husband dives head-first, and I slip down into the cold, and am at eye level with a tanker on the distant horizon. There's a strong current and, as I swim, I keep an eye on our clothes on the beach. I'm staying put, in a tussle with the water, and can hardly hold my own; my collarbones ache painfully with the cold, the marrow protesting. My legs and arms are numb. But I'm in, and that's the triumph: the liquidity all around me, the lack of firm ground, my feet floating, freed of bearing weight. The solidity of land is now observed from a distance: over the marshes and abandoned oyster beds, Margate's tower blocks stand like gappy teeth. I'm surrounded by grey: the steel water, and low-lying estuary clouds flat as an upside-down casserole dish. It feels appropriate that I've sought my sea cure so close to one of the places where this painful therapy first began, nearly three hundred years ago. Later, I'll warm up in a cafe in Margate with tea and steaming-hot fish and chips.

A few more minutes and I am stumbling out, the crisp wind almost the same temperature as the freezing sea. Then back for one more salty scuffle with the currents. A group of teenagers walking along the shore stare briefly, baffled at this middle-aged madness. Then I'm out again, cold wet fingers fumbling over swimming-costume straps. I'm bare-breasted to the estuary and feel like some modern-day Boudicca. I rub myself with a towel fast, and pull clothes on my damp, salty body.

My fingers wrapped around the mug of hot tea are yellow, but now the exhilaration kicks in, the lightness of heart, a joyfulness surging along the warmed blood vessels and tingling extremities: every cell feels as if charged with new life. There has been a ritual, a sacrifice, an offering to the waves of flesh and pain, and in return, there is restoration, life given back.

Afterwards we walk back along the tideline over the wet sand, half an eye to the ground for interesting pebbles, our ears full of the breaking waves. We pass old wooden groynes, each a sculpture, weathered to expose grain, and eroded to spiky stumps. They are a kind of silent company. Pebbles have been trapped in the timbers, and the iron bolts smear streaks of deep-orange rust on the blond wood. In some places, the groynes have been replaced by granite boulders, and their presence under the water is marked by posts topped with what look like red lampshades, the perfect place for shags to perch.

Ahead of us a flock of turnstone are standing hesitantly on the shore, just where the waves run in, and through binoculars I follow the crowd of pale-grey young, their feathers ruffled in the wind, watched over by their darker parents. As the sociable crowd move along the shore to keep away from us, the young bounce like balls of fluff. When we come too close, they take off, and the flock forms a geometric pattern as they lift above the water and swirl around to settle back on the beach behind us. The grey sunlight catches for a moment their gleaming white breasts, then they are dark silhouettes against Margate. A few seconds of movement that appear, in the exhilarating aftermath of our plunge, like a turnstone blessing on the seaside journey which lies ahead.

I

Scarborough

North Yorkshire

It was August bank holiday, and the late afternoon was already cold, with a stiff eastern breeze whipping off the North Sea, but that wasn't deterring anyone. Swimming costumes were hung on the promenade rails, a group of men peeled off wet costumes, exposing plentiful skin bright pink with cold, and they slapped each other's backs heartily in the exuberance which comes after a swim. The tide was coming in, squeezing the available beach to a few metres, and everyone was on the move, shifting deckchairs and picnic blankets until they were up against the promenade. Scarborough's tourist businesses have reason to thank the town's geography: big beaches attract the crowds, and to escape high water holidaymakers are eventually forced on to the promenade, and into the chip shops, pubs, ice-cream parlours and amusement arcades. As the sun dipped behind the town in the west, the donkey rides were going strong, the owners guiding the animals between the closely packed families. Lack of sun and a chilly breeze: this is par for the course in the English seaside resort. Yet the pleasure was evident. The squeals of excited children, the chatter of families, the sandcastles, fish and chips, and ice creams; the bustle of the seaside: memory factories.

Scarborough is Yorkshire's pre-eminent seaside resort and streams of traffic bring the crowds along the A64 from the West Yorkshire cities of Leeds and Bradford, the visitors today reflecting the diverse ethnic populations of the region. On the beach, a group of African women swathed in veils and long dresses were surrounded by windbreakers to create some privacy as they sat and picnicked, their children shivering in colourful swimming costumes. Beyond, a group of Asian women launched themselves, fully dressed, giggling and squealing, into the surf, running back and forth in the cold water, their jeans and jumpers drenched. In another group, a grandfather sat quietly over a book – it looked like scripture, and his demeanour suggested he was praying – next to his grandchildren playing in the sand. This very English habit of wet and cold is recruiting new generations and new communities, who adapt the place to suit them. As the tide receded, the glassy wet beach reflected the rosy pink of the sky in a soft gleam. Later, when I swam, the water was icy enough to take your breath away, despite the August sunshine, but green-blue and clear, every bit as exhilarating and rejuvenating as the early pioneers of sea bathing discovered on this beach in the early eighteenth century.

Scarborough – with reason – claims to have invented the seaside resort. Here, on this crumbling piece of coastline far from the political and economic centres of Britain, a revolution of the imagination began in the eighteenth century: the perception of the seaside was gradually transformed from being regarded as a place of danger, dread and dirty labour into one of healing, rest and even pleasure – of which more later. Over the course of the following century, around a hundred and fifty resorts were built on every coastline of England, and, to a lesser extent, of Wales, Scotland and Northern Ireland. No European country in the nineteenth century could match the number and size of British resorts, built to cater for the seemingly insatiable appetite for sea, sand and fresh air, although the fashion spread in time to Holland, France and Belgium, and, by the late nineteenth century, was evident in many parts of the globe. America brought further innovation in entertainment, in places such as Coney Island and Atlantic City. By the late twentieth century, widespread car ownership and cheap air travel have brought more coastlines within reach, and the world's love affair with the seaside has only intensified, encouraged by adverts of sun-drenched white sand and palm trees, from the Caribbean to the Red Sea. Meanwhile, for some, it was not just a holiday; people moved to the coast for the jobs and business opportunities as well as for pleasure. The massive shift of populations to the coast is evident across Europe and Africa, and most dramatically in the US, where the coastal population increased by 40 per cent between 1970 and 2020. John Gillis argues in *The Human Shore* that 'coasts have taken on an entirely new cultural meaning, not only for those living there but also for inlanders who are increasingly oriented towards the sea. Today we are all, in some way or another, coastal. Not only do we live on coasts but we think about them. They are part of our mythical as well as physical geography.' And the earliest beginnings lie in North Yorkshire.

Scarborough is arguably England's most beautiful resort. The castle ruins are perched high up on the cliffs above the harbour, offering a dramatic vantage point over the beaches and cliffs, both north and south, as the coastline sweeps from headland to headland. From the castle, vertiginous streets lead down to the harbour. Off one of these, an old, low-ceilinged pub provided a balcony where I paused for a drink to watch the dusk gathering over the town. Across the chimney pots and slate roofs, I looked out to sea, where a tourist boat plied back and forth across the bay. Plump herring gulls strutted along a nearby parapet, eyeing my crisps. The temperature was dropping fast – the North Sea brings a whisper of Scandinavia and the Arctic – and I pulled my summer jacket more closely around me and set off to walk briskly down to the harbour. Every age of Scarborough's history is represented on the ten-minute walk downhill, from medieval port to Georgian and Victorian seaside resort, with twentieth-century blocks of flats and modern housing estates squeezed in-between. Close to the seafront, I window-shopped, marvelling at the eccentric mix which constitutes seaside retail – wigs, damp vintage, crystals, war memorabilia and, more unusually,

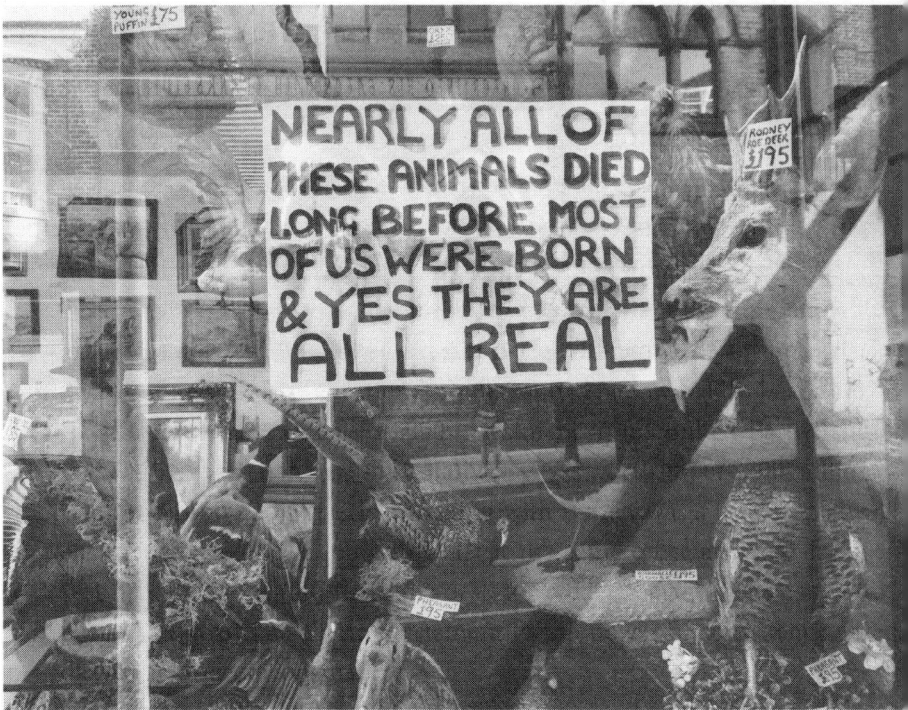

in the windows of an extraordinary taxidermist, a cornucopia of stuffed animals and birds, large and small. Handwritten notices had been stuck to the windows.

In a cafe on the front, I warmed up with fish and chips, overlooked by large nautical maps of the North Sea ringed by our neighbours: Norway, Denmark, Germany and the Netherlands. At different times this coastline has absorbed and accommodated traders, fishermen and, of course, invaders – whose dastardly deeds of pillage and looting are still a vivid memory in our cultural imagination. Vikings, Angles, Saxons and Dutch are our most intimate neighbours, as the DNA of England's east-coast population reminds us. The tousled blond and red hair of some of the children building sandcastles and playing in the waves on Scarborough's beach that afternoon can be traced to these movements of people across what was once known as the German Sea. Dozens of villages in the area carry the history of Northern European excursions to England in their names, with suffixes such as *thorpe*, *wick*, and *by*, or prefixes such as *Ug*.

The Romans discovered that the topography of the town served well for defence purposes, and built a signal station on the site of what later became a castle. The shore provided a good harbour for fishing. But its therapeutic attributes brought Scarborough fame; a spring in the cliffs was credited with medicinal waters and in the early seventeenth century it led to the invention of a new therapy: sea bathing. Two centuries followed in which Yorkshire's huge industrial wealth was invested in some of the most grandiose resort architecture in Europe: hotels, grand crescents and terraces with elaborate façades, wrought-iron railings, handsome balconies and balustrades, topped with pediments of elaborate stucco. A spa complex displayed flamboyant Art Nouveau domes. Visitors could shelter from the elements in its concert halls, and the windows of the state-of-the-art Sun Court offered protection from the wind. The Grand Hall could seat 2,000 and the Spa Theatre another 600, and the whole structure sprawled along half a mile of the seafront. The wealthy Yorkshire industrialists – the mine owners, steel magnates and mill owners – came here to rest, and to enjoy and flaunt their county's burgeoning economic fortune. An art gallery, museums, and, in due course, theatres, an orchestra and libraries sprang up: from the early nineteenth century onwards, the town had intellectual prestige as well as style. Between these grand buildings on the South Cliff, gardens were laid out with picturesque shelters and follies to enchant and delight. In 1827 the pedestrian Spa Bridge spanned the steep ravine to connect St Nicholas Cliff to South Cliff. No less than five funicular lifts were built to carry visitors up and down the cliffs (two are still in use), and the town boasted the first funicular railway in the country.

Scarborough in the nineteenth century was undoubtedly glamorous, and no building came to illustrate the town's ambition more than the vast Grand Hotel, its flamboyant towers and pinnacles visible from almost every part of the South Bay seafront. Built in 1867, it was at the time the biggest hotel in Europe and the largest brick building. It was designed on the theme of time, its four towers representing the seasons, with a floor for each month of the year, a chimney for every week of the year, and 365 bedrooms, one for each day of a non-leap year. The whole hotel structure was shaped in a *V*, in honour of Queen Victoria; the baths originally included an extra pair of taps so guests had a choice of washing in seawater or fresh. In Osbert Sitwell's novel *Before the Bombardment*, he captured the extraordinary confidence of the building:

When it was built, no other social system was deemed possible, and so it was intended like the Great Pyramid to stand through all eternity. Upon this monstrous hub, the entire system of the town's summer life revolves … In winter, the prospect gains an air of proud desolation and of great forces temporarily held in check; it compels such respect as all men must feel before some mighty machinery, mill or furnace, at rest.

The Grand is still standing, its massive bulk squarely sitting in the centre of the town, as intimidating and bombastic as ever. It was named by Historic England in 2017 as one of the top 100 places that tell the 'remarkable story of England and its impact on the world'. Bombed in the First World War, its cupolas were used to position anti-aircraft guns in the Second World War, and the building was even used for covert training for the SAS during the Iranian Embassy siege in London in 1980. But none of this weight of history helps its uncertain fortunes now. The edifice is slowly crumbling, its many roofs and balconies streaked with bird shit; they offer countless vantage points for seagulls to choose potential victims amongst the visitors eating their fish and chips and licking their ice creams on the promenade below. Grubby net curtains hang at the windows, and outside the entrance, every time I passed, a group of guests were huddled for a quick fag. Rooms are cheap but unappealing – one low-cost option had no window – but it was the history of health scares amongst guests which deterred me from booking: gastroenter-itis, several outbreaks of Norwalk virus, and E. coli in 2005.

Scarborough's glamour is over. Now it's a town of lost letters – VILLA ES LANADE and ENTRAL HOTEL – and much worse. One of its most famous sons, the disgraced entertainer and serial abuser and rapist Jimmy Savile, had a home on the seafront with his mother, and frequently turned up in the town to support good causes during his long television career. He asked to be buried in the town, placed upright so that he could enjoy the famous view, with the epitaph 'It was Good While it Lasted'. In the 1970s and 1980s Savile brought a whiff of TV glamour and national celebrity to the town, which was much appreciated locally, as Scarborough was increasingly viewed as an outdated relic of past times, perched by a grey sea, hours from a major city or mainline station. After Savile's reputation imploded in 2012, when he was accused of multiple sexual assaults of minors, the association became an embarrassment and his gravestone was dismantled. To make matters worse, allegations of abuse emerged against Savile's local friend, Peter Jaconelli, a former mayor of the town and ice-cream magnate. Questions persisted for several years about the local police's handling of complaints against both men dating back to the 1960s, and the scandal leaves a bitter aftertaste of exploitation, entitlement and abused innocence. As I walked the zigzag paths through the South Cliff gardens, Scarborough's struggle to live up to its illustrious past was evident all around. The cheap wooden handrail was broken, the flower beds were a riot of bindweed, the lawns were unkempt, and sycamore saplings proliferated.

One morning, I got up early to watch the sun rise. The sky was pale lemon, and as the sun rose over the horizon of dark sea, it burnt the water white, while colour slowly seeped into the dusky shadows of headland, beach and sleepy town. The only other people about were a few dog walkers, distant silhouettes on the gleaming beach. The tide was out. One of the best places in England to see the sun rise, I concluded, as I sat on a bench and briefly closed my eyes to the new sun's rays as they gently warmed.

'You sleeping?'

An elderly man in one of Scarborough's ubiquitous motorised

wheelchairs had pulled up alongside the bench, overweight, unshaven and cheerful. We agreed on the beauty of the sunshine and the view. He came by every day on the way to pick up his paper, he told me. I asked about the Spa, whose roofs were visible beneath us. 'It's a shame, all closed,' he said. 'There was an open-air pool that way once, but it's filled in, a shame. It's all changed.' When I asked another question, he was wary. Yorkshire-born myself, I recognised the suspicion of inquisitiveness.

'You'll be wanting a lot of knowledge,' he replied tersely. I risked one more question. 'I've got to be off,' he replied and whizzed off down the promenade.

Scarborough's decline is a story echoed on every English coastline. England may have invented the seaside resort, but by the late 1970s they were rapidly being deserted in favour of Mediterranean package holidays. The transition was swift and brutal, and the coastal resorts have struggled to find a new economic model ever since. Visitors still come in huge numbers – witness that crowded August bank-holiday beach in Scarborough – but they do not stay enough nights, or even any nights, and they do not spend enough money to sustain the resort economy. Part of the appeal of England's seaside is that the delights of sand and sea are free, so as a source of pleasure it can be cheap. Money was made in accommodation, food and entertainment. The seaside-town economy has always been seasonal, and usually precarious, even at the height of success. Hidden behind the seaside glitter and glamour lay poverty and a struggle to get by on meagre profits. Visitors came in search of escape and enchantment, and effort was always required to keep up appearances. Seaside resorts have been dominated by small, often family-run businesses, and they want their visitors to see success. With that has come a long history of discretion.

The winding streets I walked between Scarborough's castle and the harbour incorporate areas of deep deprivation. Wages are well below the national average and levels of indebtedness are amongst the highest in the country. Yet these statistics are masked by wealthier communities of well-off retirees nearby; data at local-authority or even ward level obscures the pockets of deep poverty. The most granular level of data available in the UK is known as Lower Super Output Areas (LSOAs), which cover 1,500 people or just 650 households; at this level of detail, the true extent of coastal deprivation becomes clearer. On the LSOA maps, tiny squares of deep blue mark the worst deprivation which cluster in the centres of old resorts. Inequality in seaside resorts is an intimate affair, where the prosperous and the impoverished live a few streets apart. Eight LSOAs in Scarborough are in the bottom 10 per cent of all 34,753 LSOAs in the UK; in 2017 the town had the lowest mean employee gross salary in the country, at £19,925, beating other resorts such as Blackpool and Torbay. This is part of a national pattern of low pay; of the twenty local authorities where gross salaries are the lowest, nine are coastal. Many resorts also have high levels of personal debt; in 2019 Scarborough had the second-highest rate of personal insolvencies in the country. By 2022 Blackpool had overtaken Scarborough, but it remained, along with other coastal towns such as Cleethorpes, Plymouth, Hastings and Dover, in the top ten.

Scarborough's quaint streets translate into a tragic set of health statistics on the diseases of despair: suicide in the town is 61 per cent higher than the national average, and hospital admissions for self-harm are 60 per cent higher; for alcohol-specific conditions they are 30 per cent higher, and in the under-eighteen age group rise to a shocking 116 per cent above the average. Amongst patients served by the general practice in Castle Heath in the town centre, the rate for smoking amongst adults, one of the strongest predictors of poor health outcomes, is 44 per cent, well over twice the national average of 17 per cent.

The deprivation along England's coasts leads to high levels of ill health, points out Professor Sheena Asthana, director of the Plymouth Institute of Health and Care Research. She describes an intergenerational cycle: 'Patterns are set very early by smoking in pregnancy, with higher levels of stress often due to financial issues. That impacts on the baby's birth weight. Lower rates of breastfeeding are linked to obesity and future cardiovascular disease. Higher rates of post-natal depression mean that children are then exposed to ACE (Adverse Childhood Experiences), which in turn affect cognitive development and socio-emotional health and stress. That results in under-eighteens with high levels of self-harm, hospitalisation from drug or alcohol misuse, in a circular pattern of disadvantage.'

Scarborough's story is replicated in Skegness, Great Yarmouth, Clacton, Margate, Hastings, Torbay, Weston-super-Mare, Blackpool and Morecambe. Even in prosperous resorts such as Brighton and Hove, the same issues are evident in pockets on the outlying housing estates. 'Pocket' is a term often used to describe this micro-inequality. In this national picture of seaside poverty, Scarborough doesn't rank amongst the worst; indeed on some measures it has been successful: most people are in work, unlike in the Norfolk resort of Great Yarmouth, where economic inactivity is 30 per cent – compared to the national rate of 21 per cent. Rather, Scarborough's poverty is the result of low-paid, unskilled work, and the inevitable need for loans to cover everyday emergencies such as a broken washing machine.

Educational attainment is lower than the national average in

Scarborough (Great Yarmouth's is even worse), a town where 40 per cent of working-age adults have no qualification (twice the national average), and only 9 per cent have a degree (less than a third of the national figure). The most vivid measure of this inequality is evident in life expectancy: in common with many seaside towns, Scarborough's is below the national average. If you live in a seaside town, your life will on average be shorter than if you were born inland, even in an inner city. And in many seaside towns where inequality is pronounced, life expectancy can vary widely from one end of the promenade to the other.

Some features vary but, overall, decline and deprivation are evident in every resort I visited and researched, reflecting the characteristics of these places on the edge. Situated on the periphery, their 180 degrees of sea means access to half the number of jobs, half the further education options and medical services. Geography costs them dear. Problems in these areas have been several decades in the making, and have intensified in the last fifteen years, hit by the financial crash of 2008 and austerity cuts, while the Covid-19 pandemic had had a devastating impact on the already precarious hospitality sector, the mainstay of seaside resort economies, and towns are bracing themselves to cope with the fallout from the cost of living crisis.

They are much loved, but only metres behind their seafronts are streets housing people with some of the most challenging lives in the country. Yet the pain of these communities is regarded more as an embarrassment than a provocation to outrage, indignation and political action. Unlike inner cities or post-industrial regions, the poverty of the coast is missing a narrative that can command attention. Left-wing analysis of capitalist exploitation finds little traction in these seaside economies of family businesses with marginal profits and low-wage seasonal labour (although the classic Edwardian novel, *The Ragged Trousered Philanthropists*, set in a fictionalised Hastings, portrays the exploitation of seasonal workers). Political neglect in part derives from the nature of peripherality: relations of power – economic, political and cultural – concentrate around cities inland in such a way as to exclude and marginalise other spaces.

Seaside resorts were designed as part of a pleasure periphery, their economies dependent on meeting the needs of larger, more powerful conurbations nearby; once those populations began to look elsewhere for pleasure, resorts were abandoned. They were never developed as sustainable communities in their own right, with a diverse, resilient economy. They carried the weight of expectation – and disappointment – of their visitors; their task was to live up to the promise of the adverts, marketing, hearsay and reputation.

From its beginnings in Scarborough, the seaside resort was a product of the imagination, which lives in part in the amalgam of visitors' memories, dreams, fantasies and escapism. Inlanders project on to the coastline their own preoccupations. The use of the word *line* to describe the coast's constant flux of shifting tides, sand, mud, crumbling rock and pebbles underlines how we tidy up, organise and search for certainty on this border between land and water. Our inner emotional landscape is already cluttered with expectations by the time we arrive on the beach and tuck into our fish and chips. The American novelist John Cheever defined the 'coast of the imagination', evoking the souvenirs, the cafes and pubs decorated with buoys and nets, reminders of the hard work which living beside the sea used to represent. This 'second coast' is surreal, he argues, designed to fulfil visitors' needs and desires and to sustain 'mythical geographies and historical fictions'. Here at the edge, solid land crumbles and disintegrates and its underlayers are exposed. It is unstable, and, as such, a place of possible revelation – whether that is a personal epiphany, intellectual insight or cultural revolution – as has been the case of this North Yorkshire coastline.

Scarborough's grand history began modestly in the Middle Ages as visitors came to drink the spring's waters; early in the seventeenth century those looking for relief from their ailments went a step further and braved a dip in the sea and a stroll on the beach to take the air. In 1667 Dr Robert Wittie was advising gout sufferers to bathe in the seawater, and by 1697 the traveller Celia Fiennes commented on this strange new phenomenon of sea bathing on the Yorkshire coast.

Such behaviour is now so commonplace that it is hard to glimpse how radical a departure from accepted convention this fashion represented. For centuries, a deep fear and horror of the sea and its dangerously unstable edge, the shore, was embedded in the Western tradition. It was an alien environment, argues John Gillis, who points out that in the biblical creation myth, one of God's first acts was to 'tame the unruly waters'. There is no reference to a sea in Eden, adds Alain Corbin in his stunning survey, *The Lure of the Sea*; rather the sea appears in the Old Testament as a tool of punishment in the Great Flood, reflecting the Bible's origins in agrarian society. Corbin argues that classical Greek mythology regarded the sea with comparable apprehension as full of monsters and prone to dangerous storms; the sailors of Rome and Greece hugged the coastlines. For the monastics of the Dark Ages and Middle Ages, the sea was known as the 'green desert' and presented the greatest spiritual challenge as a place to confront the terrors of sea monsters and wild storms. Shipwrecks were a common occurrence well into the nineteenth century, when Britain was still losing over a thousand ships a year, with a huge loss of life. Most ships were lost near to shore or port. The sea was that most horrifying of graveyards, snatching lives and leaving no body for burial rites.

The first sea bathers on this Yorkshire coast were regarded as intrepid as they braved the cold waves, breaking with a cultural tradition of millennia. Margate quickly caught on to this new therapy, as did a small fishing village then known as Brighthelmstone. An ambitious doctor, Richard Russell, applied his marketing acumen and codified the required treatments, and made a handsome fortune in the renamed Brighton in the process. Following Russell's tome on the benefits of sea bathing as nature's way to cleanse and defend against decay and putrefaction in 1750, other doctors followed suit, refining particular requirements for effective treatment, specifying precisely the frequency and length of immersion. The 'beach had become a medical landscape', writes Corbin; experts specified that, ideally, it should be sandy and flat, bordered by cliffs for walking and riding, and well away from the mess and detritus of a river mouth or fishing harbour. Scarborough had all these natural advantages: hard sand for walking, riding and the wheeling of bathing machines. At this point no one would have dreamt of sitting on the sand, let alone lying on it. Cliffs overlooked the natural bays and the North Sea could provide a generous supply of brisk breezes and sharp winds. Fresh air was fast developing a comparable therapeutic reputation to seawater, particularly for visitors from the polluted new industrial towns. Sunshine was far less important.

The recommended process of sea bathing was brutal, and for women tantamount to waterboarding: patients were submerged and often held under water repeatedly by attendants. Accounts testify to the terrifying nature of the experience for teenage girls who were being treated for maladies of the nerves or for gynaecological problems. To ensure modesty, bathing machines – effectively a form of carriage, invented in Margate – were wheeled down into the shallows so that women only need take a few steps before being plunged into the sea. Men were allowed more independence and might even swim. 'At the seaside, sheltered by the therapeutic alibi, a new world of sensations was growing out of the mixed pain and pleasure of sudden immersion,' comments Corbin.

Eighteenth-century visitors became fascinated by the sensory stimulation of the sea: the smell, sound, sight and feel of wind, water, sand. By the nineteenth century, the appetite for experience of the sublime – a feeling of being overwhelmed and intimidated by the force of nature – brought added intensity to the seaside visit. In 1839 Charlotte Brontë made her way to Bridlington, just south of Scarborough, and admitted in a letter ahead of her visit that 'the idea of seeing the sea – of being near it – watching its changes by sunrise, sunset, moonlight and noonday – in calm, perhaps in a storm – fills and satisfies my mind. I shall be discontented at nothing ...' When she got to Bridlington, she was 'quite overpowered, she could not speak 'til she had shed some tears. Her eyes were red and swollen, she was still trembling ... for the remainder of the day she was very quiet, subdued and exhausted.' A decade later, Anne Brontë, very ill with tuberculosis, insisted on being taken to Scarborough; her sister Charlotte wrote that Anne 'has a fixed impression that the sea-air will give her a chance of regaining strength – that chance therefore she must have'. In an age when medicine had little to offer by way of easing pain, let alone curing a disease, the pleasure of a stay at the seaside was often the only prescription. Anne enjoyed 'one of the most glorious sunsets ever witnessed,' wrote her friend Mrs Gaskell. She died in Scarborough. Her gravestone has become a place of pilgrimage, and on my visit a small pile of votive offerings had been left by devotees, including, incongruously, a small stuffed purple elephant.

Part of the paradigm shift from fear of the sea to appreciation was down to our North Sea neighbours, the Dutch, suggests Corbin. Their management of the sea through the reclamation of land and their naval mastery were reflected in a new style of Dutch painting, epitomised in the work of painters such as Jacob van Ruisdael and Jan van Goyen, in which the relationship between sea and land appeared orderly, even calming and magnificent. They demonstrated that the liminal, unstable coastal zone could be controlled and stabilised through human effort. This was a critical step in a reappraisal of the sea's *side*. Formerly, this coastal edge had provoked almost as much apprehension as the sea itself. Landslides, collapsing cliffs, the threat of flooding, gales and tidal surges made living on the coast a last resort, for the poorest communities. The better-off sited their homes well away from the edge, and sheltered from the elements. Even the sounds of wind and sea were loathed, argues Corbin, and the beach itself was regarded as unfertile, a wasteland exposed to the threat of wind and wave, useful only as a workplace for gathering shellfish and gutting fish. In *Robinson Crusoe* Daniel Defoe's protagonist builds his home inland, well away from the coast, where solid land ensured durable fortifications and the land was productive; 'the beach is the scene only of the disasters whose marks it still bears ... shipwrecks', concludes Corbin.

Over the following two centuries such was the extensive fixing of the coastline with promenades, sea walls and beach management that in the 1950s the environmentalist Rachel Carson felt the need in her classic work *The Sea Around Us* to remind readers of this inherent coastal instability: 'the boundary between sea and land is the most fleeting and transitory feature of the earth'. We had been in denial – and still are – but, as Carson points out, a beach is always on the move, as the sand and pebbles shift with every tide and storm. The unpredictability of coasts is particularly true of England's eastern seaboard. Scarborough and the Yorkshire coast have been shaped by repeated landslides. In 1993 a cliff in the town gave way, and the four-star Holbeck Hotel was left half-hanging over the precipice. Over the centuries whole villages (such as Runswick Bay and Kettleness) have tumbled down the cliffs to be swallowed up by the sea.

As the seaside grew in popularity, efforts intensified to shore up the coastal edge. Dr Russell, who did so much to promote Brighton's development from fishing village to fashionable resort, emphasised that a well-maintained resort should be 'neat and tidy' – a requirement still evident in the carefully manicured flower beds of countless seaside promenades. Tidiness – and its absence – is a recurring theme of English seaside history, a necessary reassurance to visitors that although unruly elements may be present, they are kept in order – whether that be the sea, sand, defecating seagulls or litter-dropping crowds. Part of the exhilaration provoked by the seaside's liminality is this precarious balance between order and wildness. Such was the Victorians' passion for a tidy seaside that it went too far, according to some observers. The poet Sir Edmund Gosse was outraged:

A careful municipality has studded the down with rustic seats, and has shut its dangers out with railings, has cut a winding carriage-drive round the curves of the cove down to the shore, and has planted sausage-laurels at intervals in clearings made for that aesthetic purpose. When I last saw the place, thus smartened and secured, with its hair in curling papers and its feet in patent leathers, I turned from it in anger and disgust.

The seaside, as an unsteady boundary between order and disorder, offers both refuge and prospect. We can retreat to the shelter, cafe or pub to watch the surf pounding the iron railings and granite sea wall through steamed-up windows; we can marvel at the roaring churn of an irate tide from the safety of the pier, or enjoy the vertiginous plunge of the funicular railway as it rattles down to the beach. This seaside edge licensed human ingenuity to become playful with space – the height of towers, length of piers – and in due course with velocity and gravity, continuing to indulge the appetite for novel sensation. As the decades passed, the daring of both engineers and prospective customers accelerated, leading to the heights of Blackpool Tower or, more recently, Brighton's British Airways i360, and the succession of terrifying attractions spinning bodies through the air at Blackpool Pleasure Beach.

Back in the early nineteenth century, visitors were still nervous on this unsteady ground, as Jane Austen well appreciated, making of it a pivotal incident in her novel *Persuasion*. The Cobb in Lyme Regis appears to offer a secure path, but it is deceptive, and Austen's protagonist Anne slips and hurts her ankle. To challenge such engrained anxiety about the dangers of the seaside, a contemporary of Austen, the prolific printmaker William Daniell, planned a project demonstrating the novelty of the orderly, modern coastline. Recently returned from India, where he had charted the growing empire, Daniell set out in 1813 to circumnavigate Great Britain, and make prints of its 'unaccountably neglected' coast. He published eight volumes of his prints of coastlines in *A Voyage Around Great Britain*, intent on 'illustrating the grandeur of its natural scenery, the manners

and employment of people and modes of life'. He relied on the
new work of surveying and mapping in the Ordnance Surveys:
for the first time, the coast was clearly defined, and Daniell provided a visual language for the process. He was both recording
and promoting the reinvention of the coastline.

As late as the mid-eighteenth century, rocky headlands were
considered ugly remnants of chaos, ruins of God's creation
before the Flood. The Methodist preacher John Wesley in 1743
insisted that such disordered rocky cliffs would 'melt away when
God ariseth in judgement'. Some seventy years later, Daniell
was sketching them as a spectacle to admire, and a place where
people 'gaze at, scratch their names upon a sod, and then depart,
with that fullness of satisfaction which a man ought to feel, who
is conscious that he has done all that can be done'. Land's End
had become a 'national possession', wrote W. H. Hudson in 1923,
and that was equally true for many other striking cliff formations,
from The Needles of the Isle of Wight to Durdle Door in Dorset,
and Flamborough Head just south of Scarborough. 'Coasts had
become icons of national identity and security,' writes Gillis in *The
Human Shore*. Daniell made prints of them, defining a new sense
of nation by explaining and describing its edges, in the aftermath
of the threat of Napoleonic invasion. The coastline, proposed
Daniell, was safe, and it was also both useful and beautiful. His
prints provided the proof, showing images of leisurely gentlefolk
strolling, parasol or cane in hand, across the foreshore. They gave
his customers a new appreciation of the properly framed sweep
of cliff, rocky headland and distant sea. In thousands of homes,
his prints hung on walls to inspire and educate.

Gillis suggests that 'the sea began to enter art and literature
as never before. The sea was given a cultural status. Qualities
formerly associated with the land, notably wilderness, moved
offshore [and] pristine nature found refuge in the oceans.' The

first phase of capitalism had been commercial and maritime, the second was agricultural and industrial. Land was required to be productive and profitable, so the marginal space of the coastline became a place of 'recreation both physical and spiritual'. Daniell was demonstrating through his work that this recreational edge not only offered a vantage point from which to observe the wild freedom of the sea but it was now organised: in the foreground of his prints he placed the new roads, harbours and housing which made the shoreline habitable, comfortable and easily accessible for the first time. His prints capture both the Romantic appetite for natural wonder – with a sweep of curving cliff or bay – and the colonising process whereby old fishing communities were banished, their homes pulled down and replaced by grand promenades, terraces, crescents in new property developments, and the sea corralled by new walls and jetties. The coast was changing from a zone dominated by defence, shipping and fishing – a place of work and war – to one of aesthetic experience. In Daniell's work, the fishermen and their boats were picturesque elements, the naval fleets at anchor in Portsmouth now serene.

His print of Scarborough serves as an enticing invitation. The broad road sweeps the eye from the foreground over the cliff, following the carriages, donkeys and pedestrians to the sea's edge beyond. In the distance small figures promenade on the paths which zigzag up the cliff. The coast is no longer a mess of crumbling, unstable cliff and the detritus associated with fishing boats – rotting wrecks, nets, huts – and the poverty of such marginal places. In his depiction Daniell tidies it all up and the coastline is now neat, the fences along the road underlining the importance of private property and the rule of law. Come to Scarborough, urges Daniell's print – and they did.

*

As a small child, I nearly drowned on Scarborough beach. I remember only my mother's wet woollen skirt and her legs as she stooped down to pull me out of the water. We were on a school trip, mothers had been chatting, and I had strayed into the water and lost my balance. It did nothing to dampen my delight at the visit. Other than primary-school outings, I only remember my parents taking us there once, out of season in the middle of a winter storm. My father drove our old van down the promenade as the enormous waves reared up over the sea wall to crash over the top of us, pebbles hammering down on the roof. It was terrifying and thrilling, as the water thwacked the van's side and the windscreen wipers worked furiously to provide some visibility in the grey smear of water and rain. Given the damage the salt water must have done to the van's

bodywork, I can only assume that my father's mood matched the storm.

My parents brought me up with a version of the seaside holiday that was very different from Scarborough's fun and grandeur. Every summer, they borrowed a wooden chalet on the beach in the small, picturesque village of Runswick Bay, thirty miles north of Scarborough. We have a few blurred photos, but the memories are crystal clear in my mind's eye.

My six-year-old self is lying on a camp bed in the one-roomed chalet, my sisters are in the bunks above. I can see the light fading through the window above the table where we eat; I can hear the breaking of the waves on the beach just below us through the trees. My parents are sitting on the veranda. Lights have begun to come on in the small village across the bay, where a cluster of houses climbs up the precarious cliff. I can smell my father's cigar and hear the low murmur of his voice. I am too excited to sleep. Early in the morning he will wake us to swim in the chilly grey North Sea before breakfast. Then my sister and I will race across the sands, dancing and leaping in the bracing breeze to warm our goose-pimpled flesh. Later, after careful calculations, I might spend my pocket money, saved for this annual summer spending spree, at Ices, where I have spotted a plastic pink vanity set – brushes, combs and mirror – for the princely sum of 71p. (How can I still remember the price when so many more significant memories have slipped away?) On a particular step above the lifeboat shed (each year we visited to breathe the rich varnish and new paint in the cool dark, the sense of anticipation, the vessel poised to be launched on its well-oiled ramp at any moment), a pebble in the concrete is the shape of a fairy's foot, and one morning here seagull shit lands squarely on my head. It's good luck, my sister consoles me. Later still, we will eat fish and chips at the Square Rig, and then leave our parents at the table to eat Fab ice creams for pudding and join the gang of other kids holidaying in Runswick on the Half Moon – the new concrete sea wall designed to prop up the village. If it rains, we might even go to the cinema in Whitby.

Blurred photos don't do justice to the magnificence and sense of adventure that Runswick held for a six-year-old. The sandy beach, cliffs and streams offered endless opportunities for exploration and invention. The forty-mile journey from home over the North Yorkshire moors was an hour of intense anticipation; 'Quick Sea, Quick Sea,' us five children would impatiently chant until the first one spotted the – hopefully blue – streak of sea on the horizon. In *A Sketch from the Past* Virginia Woolf asked herself, 'Why am I so incredibly romantic about Cornwall?' and concluded that the first and the most important of all her memories were those of childhood seaside holidays. 'If life has a base that it stands upon,' she wrote, it was lying in bed in the nursery, 'hearing the waves breaking, one, two, one, two, and sending a splash of water over the beach; and then breaking, one, two, one, two, behind a yellow blind. It is of hearing the blind draw its little acorn across the floor as the wind blew the blind out. It is of lying and hearing this splash and seeing this light, and feeling, it is almost impossible that I should be here.'

The historian John Walton argues in *The British Seaside* that the seaside has been the basic cultural capital for children, 'something that every child can, and should, recognise, respond to and enjoy.' It was a pervasive theme in British children's literature, and many children in the middle decades of the twentieth century learnt to read from books in which the protagonists travelled to the seaside, and fell asleep to bedtime stories of the same. Walton lists the characters who visit the seaside, from Paddington Bear, Sooty, Basil Brush, Noddy and Roland Rat, to Topsy and Tim. 'They were still coming out in the '80s and '90s at a time when

the English seaside holiday was in decline,' Walton comments, suggesting that the books reflected the nostalgia of a generation of parents. As cheap flights opened up European coasts and beyond, the great levelling, unifying childhood experience of the British seaside, evident for a century, lost its traction. Harry Potter, that spectacularly successful British export, doesn't spend much time at the seaside over the course of his many adventures, which manage to incorporate a wide range of iconic British places, from King's Cross railway station to Scottish castles and the Gothic cloisters of Gloucester cathedral.

We weren't the first generation in my family to visit Runswick. My grandfather came as a small boy before the First World War from his home in Southampton. It had always seemed a puzzling choice, when many other resorts were closer to hand, until I read in Walton's *The English Seaside Resort: A Social History 1750–1914* of how one family's preference for a 'nice, damp, inconvenient habitation that was picturesque' had taken them to the 'obscure and cheap North Yorkshire fishing villages of Runswick Bay and Sandsend in 1886 to spend their days in long walks, sketching and earnest argument'. Instantly I caught an unexpected glimpse of my great-grandmother, who was probably responsible for the family's choice, and of whom I know nothing other than her love of painting and music, and dislike of household chores. By the 1900s artists had moved into these villages' cheap cottages, and even acquired a reputation as the 'Staithes Group' as they endeavoured to catch the cool grey light of the then German Ocean. Today the cottages, still picturesque, are no longer cheap. Runswick was judged 'close to perfection' with the best beach in Britain by the *Sunday Times* in 2020. A photo of its charming white-washed cottages was chosen to front the 2019 House of Lords report on seaside towns, despite it now being largely an enclave of immensely valuable second homes.

Ninety-two-year-old Libby Mitchell has known the village all her life. Her grandparents bought a cottage in 1915 from an impoverished painter for just £5 – exceptionally cheap because prospective buyers were deterred by the recent German bombing of Scarborough. With its spectacular view of the beach and surrounding cliffs, the cottage is now worth a small fortune. Libby first visited as a small child in 1929 from her home in Leeds, and has been visiting ever since. We sat on the clifftop looking down on the bay, drinking coffee in the hotel garden, and Libby gave me an account of more than a century of family holidays – cold swims, long days on the beach, building sandcastles and damming streams, and walks along the cliffs to Kettle Ness. Even when Runswick beach was mined in the Second World War, they managed to find a safe way into the water off the rocks. Libby's holidays in the 1930s sounded indistinguishable from mine, forty years later. The family has kept leather-bound notebooks detailing visits and momentous events such as sailing races, lifeboat launches and the landslide into the back of the cottage; the entries paint a rich picture of family history anchored in a place.

'When my oldest brother, Sandy, was very ill, he asked his driver to bring him here,' said Libby. 'He drove him down to the shore and Sandy got out and stood there, tears in his eyes. Runswick was like coming home for all of us. I was one of twelve cousins and we all came here; now I'm the last one left. My father died in the 1950s, after injuries he sustained from gas poisoning in the First World War, and I always think my father's spirit is out at Kettle Ness [at the end of the bay], where he used to fish. I went back there every year until I had to give up.' Her voice falters for the first time. 'When I come here, lots of memories come back and it makes me a bit sad, but I'll be joining them all soon. I shouldn't say that, because my daughter gets upset when I talk like that.'

There is little parking in the village itself, so visitors walk down the hill, and they often choose the 'old road', which winds between the cottages, past windows looking straight into people's sitting rooms and kitchens. The village is tiny and intimate, and the stream of curious visitors makes it feel akin to a fishbowl. The village shop has long since closed, but Ices (where I bought my pink plastic vanity set), now known as Sandside Cafe, is still going strong. I watch an ancient, rusty tractor haul a boat down the beach, then it heads off to check the creels. The sea barely breaks a wave, the air is soporific. The beach is heaped a foot deep with seaweed – as it often was on my childhood holidays; one year it was so bad, tractors were brought in by the council to remove tons of the stuff rotting in the sun. It feels slimy under my unsteady feet as I gingerly cross the beach, the bladderwrack popping under my tread.

Around the time that my grandfather would travel by train to the North Yorkshire coast, the Sitwells were ensconced in Scarborough. No family better illustrates the town's glamour than these eccentric aristocrats, who bought a house there in 1870 and visited regularly over the following decades, bringing friends and acquaintances. All three of the Sitwell offspring – Edith, Sacheverell and Osbert – were prolific writers and prominent patrons during the 1920s and 1930s. Edith and Sacheverell were born in Scarborough, and all three acknowledged the formative memories of their holidays there. In January 2021 Scarborough's local news site published family photos of the Sitwells' house that had recently come to light: the drawing-room mantlepiece, armchairs and cushions draped with fabrics, pleated flounces and tasselled edges; a profusion of flower arrangements, plants and pots amidst the cluttered interiors. The double-height conservatory housed ferns and palms and exotic vegetation. The family's wealth came from the coalfields of North Derbyshire, and in 1900 the fashionable artist John Singer Sargent captured the ambience of opulence, entitlement and dysfunctional relationships in a striking family portrait: Sir George and Lady Ida Sitwell are several feet apart, no one is taking any notice of the two small boys, the father has a hand on his daughter Edith's shoulder as she stares coldly at the viewer.

Sacheverell became a travel writer, and his autobiographical *A Sketch of Scarborough Sands* acknowledges that of his many travels to study beauty 'each and every one . . . had started and had its end upon the Scarborough sands'. He recalls how on the edge of the cliff he saw 'the huge surging bay . . . a kind of immense amphitheatre in which I occupied the highest seat'. Osbert, similarly, was haunted by Scarborough's beauty: 'I gazed at the shifting and melting castles, the fulgent towers and palaces, shapes that in their turn revealed processions from the past. Vast cities washed by huge seas, that undulated in answer to the moon's call through a mist of prismatic spume.' Of the three places closest to him, one was Scarborough in winter, claimed Osbert, when 'the dashing gigantic waves batter falling cliffs, under the pounce and glitter of bitter-beaked sea gulls materialising out of the nothingness of white foam and yellow sky.'

A few years after the Sitwell children played on Scarborough sands, it became the first town in Britain to be bombed when two German battleships attacked in December 1914; eighty were wounded and eighteen killed. The raid profoundly shocked the country. The Sitwell parents were in Scarborough at the time, and Lady Ida took a piece of shrapnel to London to give to Osbert as a bizarre good-luck memento before he embarked for the front. A year later, in a famous scandal, Lady Ida, heavily indebted, was convicted of fraud and imprisoned for three months after her husband refused to settle her accounts. Osbert was summoned back from fighting to give evidence against her. The family, the town and the nation were fracturing.

After the war Osbert turned to Scarborough for the setting of his novel *Before the Bombardment.* Thinly fictionalised as Newborough, the town symbolised Edwardian opulent complacency: residents existed in 'long settled comfort and confident respectability' and their social status was marked by the 'possession of a sea view'; 'Every day, arm chairs could almost be growing, like the Empire, larger and softer.' The critic Deborah Parsons suggests that Scarborough was 'emblematic of an entropic Edwardianism basking in the face of its impending destruction'. *Before the Bombardment* is set in winter, when 'the town is empty of the fashionable and the genteel spinsters creep out of their homes'; local doctors 'aim to hasten neither cure nor death' but rather induce 'a subtle state of suspended animation which corroded the will'. The key moment occurs in the Superb Hotel – aka the Grand – where, on a quiet, misty morning in December 1914, the elderly Hester Waddington is eating breakfast in bed: 'To the people of Newborough, History had seemed dead ... [and yet] the old harridan was to rush out of her prim, neatly-kept grave, and sweep down without reason on this unoffending place.' Waddington and her bedroom 'were pulverised, fading with a swift, raucous whistling and crashing into murky air' as 'death darted at her from the sea'. The plight of the undefended seaside resort became the subject of a recruitment campaign, the novel continues, and 'Miss Waddington entered the sphere of History', her fate 'not confined to herself' but to 'many thousands of young men recruited by her death'.

I wanted to stay somewhere which gave me a taste of Scarborough's heyday, a hint of the ghosts of the Sitwells and their friends, and decided to splash out on a room in one of the town's historic hotels. It was research, I reasoned, as I eagerly browsed the hotel's spa options, and I needed to sample the full range of seaside accommodation, from luxury to tent, in the course of my journeys. But the room was barely big enough to squeeze around the bed, with a view over a car park. Worse, there was an unmistakeable smell of sewage, and the windows only opened a crack because of safety catches. When I returned to the room several hours later, the smell was pungent, and the member of staff who came to investigate agreed it was unacceptable. The duty manager followed with two large containers of lurid liquid, one green and one blue, and poured quantities of both down the offending toilet. It was the fault of Victorian plumbing, he added apologetically, but within half an hour the smell was back, and a night of broken sleep followed. In the morning, miffed by the fact that the hotel was fully booked for breakfast, and fed up with the smell, I asked to speak to the manager. Half an hour later I was told he was unavailable. Irritably, I muttered that at least I had a paragraph for my book, and set off for breakfast, by now very hungry and badly in need of a coffee. As I walked along the promenade, shouts came from behind and I turned to see a plump hotel manager running towards me waving his arms. Clearly unused to running, he begged me to 'sit down and talk' between panting breaths.

He was desperate, and, although only in his thirties, he looked as if he was about to have a heart attack as he apologised profusely. My reference to my book had been relayed, I guiltily deduced. When I returned that evening, I had a new room, a spectacular view of the sea, a bottle of white wine chilling in a bucket of ice, and tickets for complementary breakfasts for the rest of my stay. I had another paragraph.

The room was the closest I have ever come to being on a cruise ship, perched up on the cliffs with a great vista of coast-line, castle and harbour spread out around me. It felt almost precarious to be this high up, with the seagulls pattering on the roof. The drains offered no more than a faint whiff. But my sleep was broken again, this time by lurid dreams of a spaghetti-style tangle of old plumbing lying in the bowels of the hotel, stretch-ing deep into the cliffs. Unable to get back to sleep, I got up and sat by the window, while a pink-gold harvest moon rose and hung over the gentle sea, scattering its glow over a million shifts of water. As the moon slowly swung round to the south-east, it threw the cliffs and headlands into velvet black shadow, cutting the golden sea into ribbons of lace, like an eighteenth-century gentleman's lace jabot.

By morning, the beauty of the moonlit night could no longer distract me; I had picked up a stomach bug. Nauseous and weak,

I lay on my bed for the following two days, and watched the seagulls on the window ledge. From time to time, one peered into the room and yawned its gold beak wide – so wide that I could see its pink tongue. Otherwise, I listened to the cries and shouts coming up from the beach and the breaking of the waves, and watched the sunlight shift across the room. I cursed the Victorian plumbing.

When I finally felt like eating something more substantial than a dry biscuit, I knew I was on the mend. In the steamy hubbub of a busy fish restaurant, I studied the long menu of locally caught fish, scallops with everything. At the next-door table, a large man was recounting his life history, at volume, in between taking orders, and it became clear that he was the pro-prietor. Immensely plump, the buttons about to pop on the shirt straining across his belly, he described how he had once owned ten restaurants, and set up a nightclub in Swindon, with Abba as the opening act before they had made it really big; David Bowie had once performed for him. He'd had three marriages, but had always treated women with respect, he boomed, declaring, with an aplomb that silenced the whole restaurant, 'I've never had a one-night stand.' Accounts of various romantic exploits followed, and his audience at the table roared with laughter, while the rest of us listened with bated breath.

When he turned to my table, he underwent a dramatic shift of personality, and was suddenly serious. He described in vivid, loquacious detail his recent heart attack, a possible stroke, and a medical history over the last six months that included water retention, diuretics and the need for a bypass. I got lost in the sequence of tests, medication and procedures – all before I could gingerly order scallops and a glass of wine.

My stay in Scarborough left me unsteady. Nothing was quite what it seemed. Despite the calm order in Daniell's prints, and the army of property developers and promenade-builders who followed in his steps, the coast's instability, described by Carson, is tangible in the town, and it also applies to its reputation, personalities and human lives. Everything is on edge, the grandeur of the Grand Hotel in the heart of the town slowly shifting to a form of ruin.

At the far end of Runswick Bay, the eye is drawn to where the cliffs dip and then rise again in a rounded hill, a fitting full stop. This was the furthest point for our childhood adventures. We reached it by scrambling up the cliffs along muddy paths between the bracken and the bramble, the taste of blackberries and the salty air in our mouths, as the murmur of those on the beach slowly faded. We came out on the clifftop and walked along the edge of the fields of newly cut wheat, a brisk breeze coming off the sea. The promontory is known as Kettle Ness, an eerie scape of scree-covered slopes, without a scrap of green, marked by the remains of industrial workings. It is a desolate place, with only the gulls for company. The land has been quarried and hacked out over centuries of alum mining, and successive landslides have also played their part in creating these rough, scarred cliffs. The history of the village of Kettleness, only a couple of miles south of Runswick, has been as precarious. Looking down from the cliffs, remnants of buildings are visible, evidence of past landslides; a carved stone spout, characteristic of old Yorkshire farmhouses, juts out of recently tumbled earth and turf. In another place, part of an old wall is visible, a scrap of some lost field boundary or farmyard, perhaps. House martins circle over this used and lost land, darting back and forth in their magical, mesmerising dance. Up on the clifftop, the remaining houses seem forlorn, as if aware they have been abandoned by their neighbours. A large Methodist chapel stands alone in the fields, its high arched windows and steep roof designed for rousing hymns, but it has long since lost its congregation. Another relic, a railway station, is now a hostel, its past identifiable by the platforms of dressed stone cutting across the neat lawn. At the layby, a sign on a noticeboard requests information about a man from County Durham who went missing two months ago, his parked car found near Kettle Ness. The tradition of loss – of people, places and land – continues.

In 1998 a book appeared which answered the questions about this odd place which I had not known to ask as a child. In *The Floating Egg* Roger Osborne recounts how Kettle Ness played a critical role as a source of alum in the industrialisation of Britain. Alum was essential for the fixing of dyes and for other industrial processes, and from 1600 to 1870 this Yorkshire coastline was Britain's only source. In a 'dramatic story of industrial espionage, European power play and British deviousness', an enterprising Englishman broke the Papal monopoly on alum-working near Rome in the sixteenth century by hiring two Italian craftsmen to bring their knowledge to England. Cliffs stretching from Saltburn to Whitby, twenty miles north of Scarborough, with Kettle Ness at the epicentre, were ripped apart and millions of tons of rock were hacked out by pickaxe and loaded on to wheelbarrows. Every ton of alum produced another twelve tons of shale, and by the early nineteenth century 3,000 tons of alum were being produced a year: an enormous feat of excavation with nothing but hand tools.

Astonishing fossils were revealed. In 1758 a marine crocodile was found (now in the Natural History Museum), dated to 185 million years ago, and at Kettle Ness itself in 1848 a remarkable fossil reptile was unearthed, which was 7.1 metres long, its whereabouts no longer known. Lost – like so much else in Kettle Ness's history. This rich harvest of fossils provoked intense interest among local antiquarians and led to the founding in 1823 of the Whitby Literary and Philosophical Society and Whitby Museum, still going strong 200 years later. No expense was spared for Scarborough's elegant Rotunda Museum, dedicated to the new science of geology and its extraordinary finds. In the early nineteenth century this part of the North Yorkshire coast, hundreds of miles from any major city or academic institution, was a centre of intellectual inquiry and curiosity: its crumbling cliffs offered a resource library of an astonishing, and at the time deeply confusing, history. The mining and the fossil-hunting went hand in hand, explains Osborne, citing the young Louis Hunton, the son of an alum-maker, who produced a groundbreaking paper at the age of twenty-one on how fossils appeared in particular bands of rock, noting that 'zone fossils' could establish the 'temporal relations of rocks with other strata'. The context of fossils was crucial, Hunton was the first to suggest, and his insights into what is now known as biostratigraphy have been standard in geology ever since. His contribution was brief and brilliant; he died of tuberculosis shortly after.

After reading Osborne's book, I went back to visit Whitby in the late 1990s with my father, after a gap of two decades, to introduce my two small children to the delights of the Yorkshire coast. I wanted to listen to the seagulls, smell the salt and eat fish and chips. Whitby's fortunes had improved. The shabby fishing port I remembered was on its way to becoming the heritage gem which now attracts thousands of visitors a year to its steep streets, magnificent ruined abbey and quaint harbour. Famous for inspiring *Dracula*, the Irish novelist Bram Stoker's gothic horror story of a vampire, fans gather here for an annual weekend of dressing up. Our visit coincided, and my five-year-old daughter was intrigued by their Dracula-style black cloaks swirling behind them as they strode across the beach in elaborate Victorian dress and lurid make-up. The whole town was transformed into an inexplicable, fantastic carnival. My father bought me a fossil in one of the small shops which still sell the rich pickings of the area's cliffs and beaches – both halves of an ammonite, 140 million years old. I love the rounded exterior, the regular indentations of the closing spiral, believed for centuries to be snakes turned to stone by the town's medieval saint, Hilda, when she rid the town of an infestation.

Whitby is a place which breeds fantasies and mythologies. As a journalist, I travelled to Whitby to research an obscure tale of a Catholic nun who insisted she had been called by God to be a priest, and had found a renegade bishop to ordain her. The Catholic church dismissed her claim as ludicrous, but she didn't care: she had lost interest in the religious hierarchies which had governed her life and she was playing fast and loose. The interview in her cluttered kitchen was hilariously chaotic as she launched impromptu into prayer and I tried to steer her back to my questions. It was a magnificent act of defiant eccentricity and delusion, perfectly illustrated by her choice of a long black cape in which to pose for the equally bewildered photographer. Whitby is so steeped in histories, it can be hard to keep your balance.

Back in the 1970s, the chalet we borrowed in Runswick was slipping. Every summer, the fierce anticipation leading up to the holiday was tinged with anxiety that it might finally have fallen into the sea; new chunks of earth and undergrowth crumbled on to the beach each winter, and our first task on arrival was to find a path over the stream and the slippery clay up to the chalet to unload. One year, the whole hut was rolled back nine metres on logs to save it for a few more years. New walls of fortified concrete were built to prop up the village car park. The struggle continues to this day; recent shoring-up of the cliffs required considerable contributions from homeowners to ensure Runswick doesn't slip into the sea again.

In the end, it wasn't landslides which did for the chalet we used, but arsonists. Over a few years they picked off, one by one, the few dozen huts of the 1930s' development in the scrub woodland bordering the bay. Under the terms of their leases, the huts couldn't be rebuilt. I was eleven when the chalet we used was burnt, and I wrote an intense, melancholy poem in response to the devastating news. The holidays in Runswick Bay had come to an end, and they would remain perfect, untouched by cynical teenage judgement or a critical adult gaze. Like Woolf, like the Sitwell brothers, like millions of other British children, they were the foundation on which so many other memories were built.

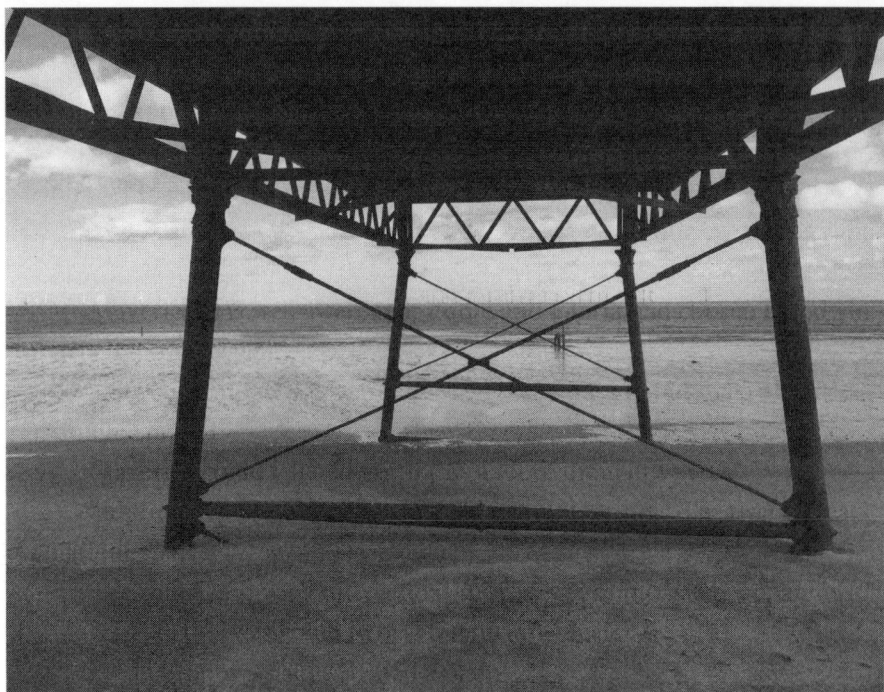

2

Skegness

Lincolnshire

After Flamborough Head's spectacular sea arches and caves, William Daniell made no new engravings on his tour of the east coast until Great Yarmouth, apart from on a detour to handsome Boston (three miles from the muddy Wash). The flat and gently undulating coastlines of the East Riding of Yorkshire, Lincolnshire and Norfolk did not meet nineteenth-century aesthetic criteria; they were not picturesque, nor did they offer the wild drama of nature or the impressive elegance of Scarborough's new terraces. Daniell saw little of interest and kept going. Huge beaches, dunes and marshland would wait until the twentieth century for their discovery by caravan-park entrepreneurs and birdwatchers. Small resorts such as Hornsea and Withernsea developed late, perched on the spit of land which hangs elegantly down, its point nudging into the Humber.

Precarity is the defining feature of this coastline. It is the fastest eroding in Europe, its combination of turbulent weather and rising sea levels eating away at the land. The annual average is a loss of four metres, but in 2019 it leapt to ten metres. One resident of the small village of Skipsea, which has been badly affected, told the *Guardian* that 'you can get up one morning and open your curtains and you've lost your fence, or your garden's gone'. At the end of one road, an abandoned amusement arcade stands opposite where Skipsea's beach social club used to be, while its caravan park loses ten football pitches on average a year. Low house prices make it hard for homeowners to move. In 2022 the head of the Environment Agency warned that whole seaside towns might need to be relocated inland over the next century because of rising sea levels. Homeowners are not entitled to compensation, and they are responsible for demolition costs when the property is judged unsafe.

The clay on this eastern coast has been slipping for thousands of years; thirty villages are believed to have been lost to the sea along the Yorkshire coast since the Middle Ages. From Yorkshire to Essex, the east coast is 'soft and vulnerable', writes Professor Robert Duck, an expert in coastal erosion at Dundee University. These adjectives are not usually associated with a coastline. YouTube videos of drone footage set to mournful music chart the process over time, showing the driveways in caravan parks which come to an abrupt end, and bungalows which are only protected by a modest length of lawn from a ragged edge of fast-approaching cliff. Empty plots have appeared between properties where homes have already been demolished; fragments of brick wall dangling over the edge mark a recent fall.

By the time I reached the rounded shoulder of east Lincolnshire, the coastline was made up of caravan parks alongside muddy marshland. The occasional cry of an oyster-catcher reminded me I was at the seaside, but the breaking surf was beyond the hummocks of vegetation riddled by muddy creeklets. The sea was barely visible on the horizon. The seals blended into the browns and pale greys. I stopped at an old coaching inn at Saltfleet, fifteen miles south of Cleethorpes, and north of Mablethorpe, Sutton-on-Sea and the implausibly named Miami Beach, so called by an enthusiast of all things American (the local caravan site has roads with names such as Fifth Avenue). Here, the absurdity of the eighteenth-century invention of a coast *line* becomes fully apparent. Does the line lie along the grass verge, the mudflats or the sandbar beyond? At what time of day or which year is the line set, given that the tides shift sand and mud and the highest tides surge to reach across the marshes and submerge the grass? On foot, I couldn't get the measure – an appropriate word – of this mudscape, since it was too treacherous to cross; frustratingly, I was pinned to the grass verge. I was surrounded by a great expanse, but it was one which couldn't be explored, and, given my modest height, offered no view.

Later at my desk, Google Earth allowed me to see from satellite height this coast's shifting pattern of earth, mud and water, and I saw how the colours – rust, cappuccino and moss green – stained and bled into each other in swirling shapes. The banks of mud were marbled by veins of water which wiggled and looped, thickening as they gained volume and resolve, until they finally reached the sea and the load of mud particles bloomed, dispersing. In the shallows, fawn sandbanks, visible even from space, undulated with their own valleys and hills in a subterranean sandscape, before the map turned abruptly to a generic – and irrelevant – blue, denoting the North Sea and the final limit of this two-kilometre-wide muddy flounce.

Lincolnshire is a county with few claims for national acknowledgement (Grantham is an exception, as the birthplace of the former British prime minister Margaret Thatcher). Wealthy in medieval times, it was largely bypassed by the Industrial Revolution. The Lincolnshire villages carry history in their names – an evocative mix of continental influences, such as Habertoft, Burgh on Bain, Gayton le Marsh, Girsby, Hagnaby and Irby in the Marsh. Its quiet villages and churches built of the distinctive greenish local stone have grown mould and moss as they gracefully age. I stopped at Theddlethorpe's ancient church, and as I parked someone appeared with a key. Visitors were rare, this conscientious churchwarden admitted. The pews had been lost to woodworm and many of the locals had lost interest. But what the county does have is caravan sites in huge number; they spread on either side of the road in serried ranks, each caravan representing a family's dreams of escape and retreat. In the smarter sites, some had elaborate decking, fencing and were decorated with hanging pots, artificial grass, rockeries, flower beds, satellite dishes and porch lights; dog bowls, barbecues, mats and steps at the ready. They were homely and yet an escape from home. Neighbours were separated from each other by a few feet and thin sheets of metal, and their windows gave on to each other's sofas and flickering TV light. I fell into conversation with one resident, Sue, out walking her dog; now retired, she described how the family had been coming here for years for the peace and quiet. The children could play outside, and as she talked it sounded like she was describing a neighbourhood from past decades – community ties, little crime and no traffic.

The district of East Lindsey (which includes Skegness) has 34,000 caravans, the highest density in Europe. The parks stir to life in March and close in November, in a seasonal pattern which echoes the movement and migration along this coast since early settlers trekked over from continental Europe. The brown sea along this coastline is the watery grave of Doggerland, an area bigger than the whole of Britain. Underwater lie the remains of a rolling fertile land, where our forebears hunted and foraged and built their dwellings.

From the beaches at Mablethorpe and Sutton-on-Sea the towns are out of sight below the dykes built to protect them from flooding. The eye is filled with the sweep of magnificent sandy beach, with only a line of water in the distance. At low tide this is the closest England comes to a desert. Mablethorpe's golden sand is firm underfoot, perfect for walking. These beaches could never be crowded, however heavy the influx of Midland holiday-makers. In this expanse of horizontals, it is the ever-changing clouds which offer drama and variety and, to assist, Mablethorpe has built a Cloud Bar for comfortable viewing.

When I finally found the sea, it was subdued, rolling between the muddy shores of north Germany, Holland and Belgium and the UK. The water was the colour of coffee. In the distance, beach and water became almost indecipherable as varying shades of brown. When the weather brightened, the light became playful, and turned the sea mauve and even shades of violet, but perhaps I had been staring too long and my eyes were playing tricks.

The cafe at Mablethorpe's North Sea Observatory offered a dry vantage point to watch the encroaching tide and the flock of small sanderlings bustling through the skim of the receding waves. Endearingly busy, their white breasts were reflected in the glassy sand, as if they each had a double in pursuit. Higher up the beach, the wind blew dry sand into gentle clouds, as if smoking. I took a mouthful of coffee, half expecting it to taste of seawater, given the similarity of colour. Next to me, my neighbours tucked into small mountains of waffle, cream and jam syrup as the rain began to slide down the glass wall of windows, and the ocean rolled closer. My husband described them as Saturday-morning waves. They weren't in a hurry, limbering across the shore before breaking lazily, as if it were hardly worth their while to collapse into surf. I missed the roar and crash of a south-coast pebble beach, the sense of an argument where land meets sea.

Lincolnshire's greatest poet, Alfred Tennyson, was born not far from Skegness, and visited the coast as a boy. The critic Harold Nicolson claims 'the greatest English poet of the sea' was originally inspired by 'the low booming of the North Sea upon the dunes; the grey clouds lowering above the wold; the moan of the night wind on the fen; the far glimmer of marsh-pools through the reeds; the cold half-light, and the gloom'. But while Tennyson may have fallen in love with the sea on the sand dunes and mudflats of the Lincolnshire coast, he chose to pursue that relationship in a quite different topography – the dramatic chalk cliffs and coves of the Isle of Wight, where his memorial was built and which are associated with his most celebrated poetry.

The Tennyson family lived in Somersby, fourteen miles from the coast, and leased a cottage between Mablethorpe and Skegness for holidays; the young Tennyson would compose poetry as he explored the shoreline. When his first volume of poetry was published, he was seventeen, and, together with his brother, he hired a coach and horses and they drove to Mablethorpe 'to share their triumph with the wind and the waves'. It was a rare moment of happiness in an unhappy childhood: one of twelve children, his family was dogged by tragedy, with multiple mental breakdowns, violence, substance abuse and the shame of epilepsy (then associated with sexual excess). His poem 'Mablethorpe' reflects the misery of his early life in Lincolnshire:

> *How often, when a child I lay reclined,*
> *I took delight in this fair strand and free:*

Here stood the infant Ilion of my mind,
And here the Grecian ships did seem to be.
And here again I come and only find
The drain-cut levels of the marshy lea,
Gray sandbanks and pale sunsets, dreary wind,
Dim shores, dense rains, and heavy clouded sea.

Yet tho' perchance no tract of earth have more
Unlikeness to the fair Ionian plain,
I love the place that I have loved before,
I love the rolling cloud, the flying rain,
The brown sea lapsing back with sullen roar
To travel leagues before he comes again,
The misty desert of the houseless shore,
The phantom circle of the moaning main.

His grim depiction would be familiar to a visitor on a sodden bank holiday, although they might not share his dogged affection. Tennyson frequented the southern part of Skegness, where it spreads to dunes for a few miles before reaching the salt marshes of Gibraltar Point. It was easy to imagine the young poet whispering verses amongst these marshes on the day we visited. A brilliant early-spring sunshine glistened on the slabs of creased, sagging mud left behind by the receding tide. Old boats and wooden pontoons creaked in the heat, slowly rotting where they had been abandoned. Silvery pools of shallow water reflected the cloudy skies like a mirror, and the land beyond floated in a thin strip against the sea. All seemed insubstantial, nothing but reflection and moisture. In the distance, seals stretched out to sunbathe. The air vibrated with the sound of larks, a reed bunting flitted between the winter-bleached reeds, and, closer to hand, the handsome, black-necked Brent geese murmured conspiratorially.

But, unlike Wordsworth in the Lake District or Thomas Hardy in Dorset, Tennyson's legacy has not shaped a national affection for the Lincolnshire coast. The Isle of Wight has muscled in to claim him. For the heritage-led regeneration projects beloved of coastal resorts, Tennyson has not proved of much help in relaunching Mablethorpe or Skegness, unlike the way Margate has used J. M. W. Turner as the centrepiece of its cultural regeneration, with the building of the Turner Contemporary art gallery. The Lincolnshire coast has never been romanticised; cultural prestige accumulated on the dramatic Atlantic shorelines of Cornwall and West Scotland, which met the influential eighteenth-century philosopher Edmund Burke's criteria for the sublime as overpowering and awe-inspiring. In contrast Lincolnshire's flat fields of potatoes and cabbages were too useful and mundane to offer a Romantic counterpoint to the industrial and urban. Appreciation has been in short supply, as with the low-lying shorelines of Essex and Lancashire. Only in 2019 did Lincolnshire finally find some of the affection which has been lavished on virtually every other English coastline, in a book by the art critic Laura Cumming, *On Chapel Sands*: 'The character of Lincolnshire as it meets the sea is level and low; a great plane of clay-ploughed fields and bare-branched willows that spread into the distance like a Dutch winter landscape. Every modest haystack and spire seems a mile high as you pass beneath the wheeling arc of bright sky.'

Full of enthusiasm for the area where her mother grew up, Cumming saw how great artists could have shared her affection if they had ever got there.

> On a clear day the sea at Chapel is like a Seurat, crystal clear in its frozen shimmer of sand-sea colours. On a breezy day it is a Turner, the waves meeting the sky in a rolling vortex of

liquid and air, the two elements blending in one of his rapid watercolours ... The sweep of sand stretches away in that blurry miasma of motion, colour and light ... [Turner] could be right there at work on Chapel Sands ... the sea remote and withdrawn, a distant bar of blue far away across the strand ... even more surprising than the flatness is the way the sand appears to merge with the sea ... On a still day they become one vast continuous expanse, an optical illusion only dispelled when a chink of reflected blue sky spangles the water or a sudden gust troubles the surface.

Seen through Cumming's eyes, the Lincolnshire coast becomes magical, its emptiness a soothing balm for exhausted twenty-first-century screen-dazed eyes; finally, Mablethorpe, Chapel St Leonards and Skegness have a book-length love letter.

'We go every year, it's like home from home,' Cumming told me. 'It's so cut off now, isolated and unvisited. It's as if it has drifted away from the east of England. But people miss it. After my book came out, lots of people wrote to me explaining how they had holidayed in Skegness for years and remembered it with total joy, but had then stopped going. During my research I came across people who had visited every year for decades. I like the fact that it has no "Southwold aspect" [the Suffolk resort popular with the north London middle classes], no media types.'

The focus of Cumming's book was the bizarre kidnap of her three-year-old mother from the Chapel St Leonards beach. She was found a few weeks later, in good health and in a new set of clothes; Cumming set out to unpick the strange secrets which enmeshed her mother's early life growing up in Chapel and Skegness. The kidnap was hard to understand, writes Cumming, 'there are no coves, dunes or rocks where an adult could hide a child; everything stands in open view'. She pieces together a story of illegitimacy, adoption and shame in a small, tightly knit community, and how her mother's origins were an open secret among neighbours. The account finds an echo in the landscape and the misleading assumption that in somewhere so open and spread out, everything should be clearly visible. It resonated with this coastline's older history of loss. Important things, such as children and land, can vanish into thin air. The apparent monotony of this coast was quietly deceptive.

Skeggy is a great, spectacular eruption amidst the modesty of Lincolnshire's landscape: a sprawling brash beast of a seaside resort, drawing millions of inlanders from the East Midland cities of Nottingham, Leicester and Derby. Along the A158 route east, the land is made up of flat fields of cabbages criss-crossed by ditches. As the trees dwindle, and the miles speed past, the more improbable it seems that there will be anything at the end of the journey. The March landscape provides a limited palette of muted greys and browns until we arrive, blinking, in the neon-lit exuberance of Skegness. Attractions are crowded on to the seafront – Aladdin, Wild River, Pleasure Beach, Oasis Bingo, Jolly Roger Boating Lake, Tenpin Bowling, Suncastle Indoor Adventure – and block the view of the sea. The town's population of 21,000 swells by more than 100 per cent in the holiday season, and is dwarfed by its four million visitors a year. But a large proportion are day trippers with a close eye on their budget; the holiday-resort economy was always predicated on overnight stays and big spenders, and Skegness has struggled to attract either in recent decades. The town has a raw, unfinished appearance, its edges petering out in dishevelled car parks and shabby hotels.

In 1873 Skegness was a small fishing village, a relative latecomer to the resort phenomenon, but quick to catch up once the railway line had been built. By 1881 it had a pier and pleasure gardens, and two years later steamboats and bathing pools. The landowner, the ninth Earl of Scarbrough, donated land for a church and chapels, a school and a cricket ground: Skegness was to be respectable. By the August bank holiday of 1882, the number of visitors had reached 20,000, and they continued to climb steeply: by 1913, 750,000 visited annually a town that had a resident population of under 4,000. Its rapid expansion was an outcome of the legislation creating bank holidays, of rising household income, and the railways. A new era of mass leisure had arrived. Skegness made a virtue of necessity by redefining the North Sea's cold winds as an invigorating attraction for the dwellers of smog-ridden industrial cities. 'Skegness is SO bracing' became the famous tag line for

the 1908 advert of the Jolly Fisherman, a rotund man blown along the beach, still popular today on signs and postcards. By the late nineteenth century, it was known as the Garden City by the Sea, 'a champagne bath to re-energise the body, and champagne air to fill the lungs'. Between the wars, the town council developed new entertainments to keep up with other resorts in East Anglia and Essex.

Esplanades, a boating lake, the Fairy Dell Paddling Pool and the Embassy Ballroom were all opened in the 1920s. Laura Cumming's mother visited as a small girl in the 1930s and remembered Skegness as 'lively and elegant', even sophisticated, with smart hotels which catered for the Midlands' middle class, such as the Chatsworth, Park Lane and the Savoy. The town had a major department store, theatre, cinemas, an excellent grammar school and library. Even in the 1960s, the Lincolnshire resorts displayed some vestige of their earlier aspirations; Cumming still has a bill for one of her family's annual stays at the Vine Hotel in Chapel when she was a child. 'It's written in a very elegant copperplate handwriting. Refined. But the prices were cheap. Now it's a pit.' When I visited the bar, it was festooned with the St George flag and a banner reading, 'Your body is not a temple but an amusement park.'

In March 2020 it was hard to see glimpses of Skegness's elegant past. The main street was dominated by charity shops and the department store's windows were covered with posters for a closing-down sale. Huge pubs on the seafront advertised cheap beer. Skegness has an unenviable reputation; when I googled the town, top of the 'People also ask' prompts was: 'Is Skegness full of chavs?' A reminder of how England's relationship with its seaside resorts is tangled in a nasty and condescending argument about class. But on the walls of the cafe in the town centre where I had fish and chips, I found framed black-and-white photos of a lost heyday: a park with neat paths, tidy lawns and couples promenading in their best hats and high heels; a bridge with a handsome balustrade bestriding a waterway where a boat passed laden with passengers; the Fairy Dell, where rustic fences added quaint appeal amidst the trees and a small child in white ankle socks and a cotton bonnet held tight to her nanny's hand. Ironically, Skegness's elegant, continental-style park and exotic gardens with palms and ruined follies incubated the taste for the foreign which would presage fierce competition in the future, as 'continental' – long a term of suspicion – shifted to an association with beauty, pleasure, fantasy and adventure.

Skegness's greatest invention of all was the work of a Canadian fairground hand. In 1929 this enterprising beach stallholder, Billy Butlin, took a gamble and built amusements on the seafront; he developed a carnival in 1933, and in 1936, at Ingoldmells, just north of the town, he launched his first holiday camp. His empire grew to ten camps at its peak, and Skegness is still the flagship. On a cold Friday afternoon in March, a long stream of cars queued up at the gate to enter the huge complex. There were few clues of the delights from the outside: the vast windowless buildings could have been a large logistics complex. Butlin's has outgrown Skegness, and few of its visitors who stay at the cute apartments – for a packed schedule of music, comedy and magic, pools, rides and restaurants – bother to visit the town. They probably don't spend much time on the windy beach either. A roster of children's entertainment ensures that visitors have no reason or time to leave until check out; Butlin's camps are happiness factories and anywhere places, with almost no connection to their physical environment beyond the road that takes the visitor there. At Ingoldmells, the curving metal frames of the roller coaster rear high over a sea of flat caravan roofs which stretch in every direction. The medieval parish church squats, mildewed, in the old heart of the village, bewildered by the surrounding fields' crop of metal boxes.

By the time Butlin's opened in Skegness, holiday camps had become well established, often run by Christian organisations, churches or trade unions; accommodation was simple and cheap, and the camps offered the health benefits of fresh air and sea-water alongside the character-building experience of camping camaraderie. The Scouts, founded in 1908, and church boys' clubs inculcated such tastes young. Butlin's took the concept and made it fun. The aim was to keep prices low, but incorporate a new level of comfort plus plenty of entertainment, with amusement parks, dance halls, restaurants, tennis courts and playgrounds. Butlin's formula was timely. Two years after it opened, in 1938, a parliamentary act brought in paid holidays, and 15 million workers were now entitled to a week's leave in the summer. Butlin's became hugely popular and, in time, inspired competitors. An international name, on occasion it shipped attractions to the continent. In one of the most tantalising chapters of the company's history, Butlin retrieved 400 tons of big dippers and dodgems from the international fair at Liège weeks before the Nazi invasion of Belgium, by hiring canal boats to ferry the equipment by night to the coast and thence back to Hull. Butlin's camps ensured that England was at the vanguard of another wave of invention of the seaside resort, which carried them through a golden age in the post-war period, when the experience came within the reach of almost every family.

Part of Butlin's success was recognising that the English seaside visitor needed to be kept busy. Shortly after he arrived from Canada in the 1920s, Butlin visited the resort of Barry Island in Wales, where he saw how the boarding houses turned their visitors out straight after breakfast, regardless of the weather. Given the vagaries of the British climate, these visitors needed something to do, he concluded, and, with his background as a fairground hand, he had plenty of ideas. One of his most fabulous was to bring an elephant to Skegness beach, a sight which must have had spectators rubbing their eyes with disbelief. Visitors' expectations of comfort and entertainment were rising, and a windy beach was no longer enough. Another insight was to prove important: holidaymakers needed to be put at their ease, even jollied along. At a holiday camp in Canada, Butlin had been impressed by the friendliness, and attributed it to the fact that everyone had paid the same. Noting that the deep class divisions of England were reflected in how its resorts had developed, Butlin spotted that in this new era of mass leisure, millions were coming to the seaside who were uncertain how to behave in this unfamiliar context. Butlin imported Canadian bonhomie and friendliness to put people at their ease, and Butlin's attendants – they came to be known as Redcoats – were instructed to set the tone, cracking jokes and bantering with customers as they marshalled them through the day's activities. Butlin wanted people to enjoy themselves, and he didn't trust them to know how to do that without some help.

That insight now seems patronising, but it is backed up by the findings of the famous social research project started in the 1930s, the Mass Observation survey (MO), when its volunteers visited Blackpool. The researchers fanned out across the town to observe and interview visitors, and they were worried by what they saw. On the packed beaches, 'the crowd was less friendly and joyful than it should be', one noted, adding that 'people tended to stay in their original groups'. They were surprised by 'how little laughter there was in Blackpool crowds' and were concerned that instead, people were 'too engrossed in pennies, resting, reading, gazing and gaping'. The MO researchers were particularly surprised to discover that visitors had very little interest in the sea. In their interviews with sixty holidaymakers, only twenty-eight confessed to even liking the sea. In his *English Journey*, published in 1934, J. B. Priestley came to the same conclusion, and in R. C. Sherriff's iconic novel about one family's holiday in Bognor Regis, *The Fortnight in September* (1931), the mother's attempts to conceal her dislike of the water is a persistent theme. A Butlin's holiday camp, away from the possible discomforts of sand and sea, ensured that such reservations were catered for.

In 1936 the poet W. H. Auden offered advice on this new fashion of holidaymaking for those unsure about how they were supposed to enjoy themselves. In *The Way to the Sea*, a publicity film to mark the electrification of a railway line to Portsmouth, then promoting itself as a resort, Auden urged:

> *Be extravagant*
> *Be lucky*
> *Be clairvoyant*
> *Be amazing*
> *Be a sport or an angel*
> *Imagine yourself as a courtier or as a queen.*
> *Accept your freedom.*

Here was a new set of aspirations, the precise opposite of the working-class habits that chapel and factory master had vigorously promoted for the previous century, requiring a re-orientation away from a culture of hard work, thrift, modesty, caution, where collective endeavour was privileged over individual achievement. Auden was encouraging the new consumerism coming from America via cinema screens, and was even advocating gambling, fortune telling, and displaying one's sporting prowess or beauty in competitions. The seaside had become a realm of fantasy and sensation, a stage on which to show off and, above all, to *indulge*. It had become symbolic of personal freedom, and the implication – even Auden couldn't be explicit – was that these freedoms also applied to sexuality. In the 1930s visiting the seaside was to visit the future.

That only exacerbated the anxiety about how to behave. This was a brave new world, and Butlin offered reassurance and guidance in the form of structured activities. He understood his customer, and his business boomed, as did that of his competitors, Warner's (their first camp opened in 1932 on Hayling Island) and later Pontins (established in 1946), also catering for this huge market. Those who didn't opt for the holiday-camp experience were left to navigate nervously the class tensions, which were evident well into the 1960s, as a short documentary about a Skegness hotelier, Cyril, beautifully portrayed in 1966. Cyril confided to the interviewer that he needed to 'break down' his guests to ensure they had a good time. To illustrate his point, the camera panned across the hotel's small, crowded dining room as families waited in silence for their lunch, not a smile amongst them. As Cyril served lunch, he kept up a cheery stream of repartee, questions about the weather, and what guests had been doing, and what they were planning to do that afternoon. Judging by their awkward manner and stilted replies, Cyril's efforts were not effective, or perhaps the presence of a film camera was adding to the embarrassment. Lunch in a hotel was still an exacting social performance, and in this documentary guests were smartly dressed in suits, ties and dresses. The English were awkward on holiday, uncertain of the etiquette of mixing with other social classes and how to maintain their status.

Cyril's preoccupation during the sixteen-week Skegness season, much like Butlin's, was to entertain the visitors, and the camera followed his hectic days of not only running his hotel but also compering donkey races, talent shows and amateur dramatics. At the end of this marathon, Cyril admitted he went on holiday to Majorca, but that salient fact did not undermine the commentator's confidence in Skegness's future as a holiday resort; he propounded that it 'is a fantasy town in which visitors can involve themselves in magical illusions, forgetting tedium and the worries of life back home'. The footage told another story: everywhere the camera landed, people were sleeping, mouths often gaping open, on lawns, terraces or beaches, in deckchairs, or under newspapers with waistcoats loosened.

Skegness was where my husband spent his holidays as a small child. In front of my keyboard, I spread out the black-and-white photos that my mother-in-law has sent of my husband, aged

between two and five, on his annual holidays to Lincolnshire. The family were following tradition. My husband's grandfather had had a caravan at Skegness for decades, and visited regularly from his home in Nottingham. In the late 1960s my husband, blond-haired with a toothy grin, was riding his donkey with relish; in another photo he is cheekily driving his bus on the merry-go-round, and, in another, emptying buckets of water with a child's intense concentration, Skeggy pier in the distance. The paddling pool was surrounded by a crowd of mothers in headscarves and sensible skirts, and dads with their trousers rolled up. Despite the sun, everyone was wearing jumpers in the bracing breezes. In time, my in-laws bought a car, and holidays moved further afield to Devon and Cornwall.

Nostalgic and curious, my husband accompanied me on the research trip to Skegness, and we decided, for old times' sake, to stay in a caravan. Having passed thousands of them in the course of my journey down the Lincolnshire coast, I was intrigued. The caravan park entrance was flanked by two large artificial water-falls, and the roads were lined with tall, neatly clipped evergreen hedges; it was a town in itself, with its own leisure centre, shop and cafe. Since the Skegness seafront was a half hour's walk away through tired suburbs, some visitors probably stayed put for a good part of their stay.

It was early in the season and by the time we arrived we had been upgraded to deluxe. Our caravan was considerably more comfortable than the one my husband would have stayed in as a very small child before the family upgraded to a chalet, or the one which my family used in the 1970s. We had unimaginable luxuries such as central heating, two tiny bathrooms and an elab-orate decor of beige, glass and brass. Big windows looked out on to a pond, but the caravan still had an unnerving unsteadiness – it shook when my husband laughed. Everything had its place, so I felt like I was in a well-stocked cutlery drawer with dinky plastic dividers; bodies just fitted into the small shower and bedroom with a few inches to spare. After we closed the curtains, we could hide from the neighbouring caravans, three metres away on either side, and listen to mallard ducks squabbling amongst the reeds. In fact, for the three days we stayed, we barely saw another soul on the huge caravan site; the headlines were full of anxiety about the looming pandemic.

Our first task was to see if we could find the chalets where my husband had stayed as a small boy. Might they jog early memories? With guidance from my father-in-law on the phone, we found ourselves on the outskirts of Sutton-on-Sea. The chalets had fallen on hard times. Large bins were lined up outside every front door, and yet the concrete yard was strewn with rubbish. The chalets appeared now to be permanent homes rather than holiday lets, but it was midday and there was no sign of life. They crouched below a nine-metre-high sea wall which blocked any view of the neighbouring beach. Across the road, a derelict house carried the marks of its past history, with tattered signs advertising bed and breakfast and a carvery every Sunday. The string of small resorts north of Skegness – Mablethorpe, Chapel St Leonards and Sutton-on-Sea – had a brave air of just hanging on. They were haunted by the ghosts of closed businesses, with faded signs advertising the failed ambition and lost hope. But the flower beds were carefully planted and the sea wall at Mablethorpe was decorated with brightly coloured art works by a local group from the University of the Third Age.

In Paul Theroux's *The Kingdom by the Sea*, his criticism of this coastline was scathing: 'nothing is more bewildering to a foreigner than a nation's pleasures and I never felt more alien in Britain than when I was watching people enjoying their sort of seaside vacation'. Mablethorpe, he continued, was 'thronged with shivering vacationers, the coast of last resorts ... no more fun than a day out on the prison farm'. He was even more cruel about Skegness: 'it deserved its ragged-sounding nick-name. It was a low, loud, faded seaside resort. It was utterly joyless. Its vulgarity was uninteresting. It was painfully ugly. It made the English seem dangerous.'

Skegness is the one resort on this coastline which is not divided from its beach by high dykes, but the beach lies the other side of a sequence of fairgrounds, amusements, aquarium, bowling alleys and stalls offering doughnuts, chips and sweets. Workmen were painting shutters, tidying up the arcades for the new season, but Covid-19 was looming, and with it the unimaginable possibility of a summer season with no visitors and the vital £450 million revenue on which the town depends.

A few months later, Skegness was identified as second only to Cornwall's Newquay as the town most affected by Covid, with the highest proportion (55 per cent) of employment in locked-down sectors such as hospitality.

Colourful cartoons along the side of one building offered a gently saucy humour: a lady whose bikini top barely covered her bulging breasts declared, 'Anything you say will be taken down,' and her comparably equipped friend replied, 'Knickers,' over the head of a man sweating profusely as he ate his fish and chips. In another, 'Do you think the doctor could give me any pills to increase my sex urge?' asked one knobbly-kneed punter with a knotted handkerchief on his head. 'What happens when you get scared half to death twice?' asked another. The cartoons' recurrent themes were marital irritability, embarrassing bodily ailments, obesity, burps, farts and piles, and old men ogling breasts. The old-fashioned humour dated from an era when sexual harassment was routinely dismissed as a bit of fun.

Skegness's story of boom and bust has been spectacular; since the 1970s its geography worked against it, and cuts to its railway services accelerated the decline. It has no direct train from London (the journey takes three hours – as long as getting to Paris), while Nottingham is over two hours away by rail, slightly faster by road. It trades on being a 'traditional' seaside resort, but the forced jollity cannot mask the air of dilapidation; seafront hotels which have seen better days, shuttered shops, and the once-elegant municipal gardens now running to seed. The pier was badly damaged by fire in the 1990s, and the remaining black shafts stalk across the swathe of sand to a distant tideline; when we visited, the sea was so far out we never reached it. Attractions still pump out music, and their neon and glitter endeavour to draw in punters. Perhaps in peak season the crowds make it more convincing, but in March a version of

Tennyson's melancholy was hard to shake off. In the cafe where we ate our fish and chips (a snip at £4.95 per person), there was one other customer, an elderly lady who ate her lunch with slow deliberation. A beautiful, middle-aged waitress watched her three customers closely from the back of the cafe, and when we moved to leave, she came forward to hold the door open; the service in Skegness was second to none.

Next door, Chapel St Leonards was also struggling. The town magazine had adverts for a Citizens Advice Bureau, drop-in coffee mornings to combat loneliness, and exercise classes for the aged. In common with the neighbouring resort towns, mobility scooters were ubiquitous. It was chilling to think that the nearest hospital was in Lincoln, a two-hour bus ride away or a forty-mile ambulance trip. On a sea wall, someone had gone to the trouble of cementing a small heart-shaped slate plaque dedicated to the memory of their two-year-old dog. At least Chapel had those long windswept beaches in its favour, providing space for any size of grief.

Over 80 per cent of the residents of Skegness and Mablethorpe live in areas categorised as amongst the 20 per cent most deprived in England. Around a third of residents have no or low qualifications, and the rate of economic inactivity – either due to retirement, disability or care responsibilities – is 42 per cent of the population, compared to the national figure of 21 per cent, which leads to low levels of disposable income amongst residents and a high number of small businesses that were considered precarious even before the pandemic. Employment is predominantly low-skill and seasonal. Young people have limited access to further education and training, given the long travel times from the coast. The broadband and mobile phone coverage is also limited, further reducing the options for diversifying the economy away from tourism. In 2017 East Lindsey was cited as the area with the third highest level of anti-depressant pre-scriptions in the country. Only Blackpool and Sunderland had higher rates. Interviewed by news reporters about this coastal pattern of mental ill health, Dr Jay Watts, a consultant clinical psychologist, speculated that it was due in part to the fact that inhabitants of seaside towns were 'surrounded with the ghosts of a better time'.

The demographic profile is steadily ageing; in the primary-care network of general practitioners in Skegness and Mablethorpe, 32 per cent of patients are over sixty-five, com-pared to 18 per cent nationally. The number of young people is falling as youngsters move away in search of jobs and training, while the over-sixty-fives are increasing as more retirees arrive; between 2011 and 2019 there was a nearly 20 per cent increase in the elderly; the ratio of population movement was one young person fewer for every additional 3.7 older people. In the next twenty years, the sixty-fives and over will increase by 43.9 per cent, and the over eighty-fives by a staggering 116.6 per cent. With the imbalance between those of working age and the elderly, the shortage of care workers is already apparent and, given the projections, will reach crisis point. Lincolnshire has one of the highest levels of unpaid care in the country, with a quarter of carers providing fifty or more hours of care a week, according to the Chief Medical Officer's 2021 report. The private care sector employs 20,000 workers, and is one of the biggest employers in the area (further reinforcing the problem of low pay). Within fifteen years, there will not be enough working-age people to provide the care needed in the area.

The ageing demographic and high levels of economic inactiv-ity lead to an endemic problem of loneliness and social isolation. The area was awarded a £2.7 million National Lottery Fund grant in 2015 to run a six-year project on befriending and friendship groups. Hence the noticeboards in places like Mablethorpe and Chapel St Leonards with their offers of coffee mornings and keep-fit. A common pattern is that retiree couples with modest incomes move to this coast after a history of holidays here, and after one dies the other is left alone, far from the family support and social networks of their original town.

Use of the NHS increases with age, but this is not always fully reflected in the formula for NHS funding and puts additional pressure on budgets in a place like East Lindsey. This is exacerbated by the large seasonal influx of temporary caravan residents, which the NHS has estimated adds an extra £22 million of health costs a year in the area.

Life expectancy drops dramatically in the least prosperous areas along the coast. For men it is 10.3 years less than in wealthier areas of the county; for women the figure is 7.2 years. East Lindsey has a higher premature mortality rate compared to the rest of Lincolnshire, and the highest emergency in-patient admissions in the country, and the rate is increasing. Research indicates that emergency admissions can indicate insufficient management of long-term chronic conditions amongst the elderly, or evidence of a 'significant health-service deficit'. There is a high prevalence of cancer, asthma, chronic obstructive pulmonary disease, arthritis, dementia and all types of cardiovascular disease. East Lindsey illustrates an issue which only recently has begun to be recognised by policymakers: people living in coastal communities are more likely to have these long-term life-limiting health conditions, and they are more likely to die earlier of them. The health-service deficit is in part a reflection of the relatively sparse population, which means a journey is usually needed to access health care. It is also indicative of the difficulty of recruiting health professionals, a problem which

East Lindsey shares with other coastal towns. Health Education England acknowledges that 'despite coastal communities having an older and more deprived population, they have 15 per cent fewer postgraduate medical trainees, and fewer consultants and 7 per cent fewer nurses per patient'. In a tragic inversion of the seaside resort as therapeutic, these towns are now struggling with a badly neglected public-health challenge. The Chief Medical Officer's report in 2021 warned that 'if we do not tackle the health problems of coastal communities vigorously and systematically there will be a long tail of preventable ill health which will get worse as current populations age'.

In villages such as Sutton-on-Sea and Chapel St Leonards, the history of flooding is too ancient and vivid to risk building on top of the dunes, so they hunker down in the lee of their massive sea defences. The verse of Jean Ingelow, a Victorian poet much admired by Tennyson, was once a staple of children's poetry anthologies, and my mother used to read 'The High Tide on the Coast of Lincolnshire' to me when I was a child. I loved this tragic poem about the beautiful Elizabeth calling to gather her cows on the Lincolnshire marshes, her two small children at her side, as Mablethorpe's bells rang to warn that the tide had broken through the sea wall. All three are drowned, her husband only saved by taking refuge on his cottage roof. Ingelow was born in Boston, south of Skegness, and her poem drew on local knowledge of the floods which have been a part of everyday life in the area for centuries. Skegness was flooded in the sixteenth century, and had to be rebuilt further inland; the first sea defences were erected in the medieval times. Its development in the nineteenth century required new sea walls north of the town, and these were eventually extended for fifteen miles to Mablethorpe.

In the infamous North Sea floods of 1953, the tide broke through in several places and claimed forty-two lives in the area. Lincolnshire's defences held against a bigger surge in 2013, but the topography is not conducive to complacency. The most recent report by the government's Environment Agency concludes: 'Over half the normal high tides are above the level of the land behind the defences and without them floodwater would reach up to fifteen kilometres inland regularly, and the frequency of inundation would make the land uninhabitable.' To illustrate the point, the report included a schematic cross section showing a sea level which drowned multiple villages, marked by their church spires, reaching a new coastline at the Lincolnshire Wolds, twenty or so miles inland.

Predicted sea-level rise and an increase in winter storms aggravate ancient anxieties. On the online national Flood Risk Map produced by the Environment Agency, mauve denotes the highest level of risk; as you scroll over England, the scattered mauve fragments litter the country, interspersed with the thick orange lines which mark flood defences. In most places it's a marbling of England, but there is one solid block of mauve which covers the Lincolnshire coast, submerging Skegness, Mablethorpe, Sutton-on-Sea, Chapel St Leonards, and dissolves the outline of the Wash so that the sea licks at Peterborough's outskirts, flooding the fens and Boston.

To offset that risk, millions have been spent every year since the 1990s on the Lincolnshire coast. Sand is pumped out of the seabed and brought onshore to spread along the beaches; 400,000 cubic metres of sand were moved in 2020 alone, in a process known as beach nourishment. Between Mablethorpe and Skegness this elaborate annual feeding of the beach with new sand is designed to help it with the pummelling of heavy waves – the shallow gradient of the beach absorbs the wave energy and limits their depth, thus protecting the harder sea defences higher up the shoreline, such as sea walls.

Along much of its eastern and southern coastlines, England has been trying to keep its beaches in place; it has been a Sisyphean task, which has only grown more complicated and expensive. Wooden groynes which counteract the currents pushing pebbles and sand along the shore mark countless beaches along the East Anglian and south coasts. In other places the huge granite boulders of 'rock armour' have been imported from Norway, and large parts of Essex are protected by concrete walls. In Lincolnshire the Environment Agency has warned that beach nourishment is no longer sufficient, and one option is lines of rock armour interrupting the expansive vistas of sand.

The Environment Agency report concludes that 'coastal communities must take ownership of their risk of flooding and build a better understanding of what the risk may look like in the future'. But I'm not sure what 'ownership of risk' of devastating coastal floods means: constant anxiety? Stoicism? Tracking winter tides obsessively? Life in Mablethorpe is already a struggle. As a reminder of its mournful bravery, I pinned above my desk a photograph I took in the town: a sign for a closed amusement arcade showing a cartoon portrait of a sailor boy with a crazed grin: 'Jacksons of Mablethorpe, 1925'.

The meeting point between land and water along this coastline has shifted dramatically over thousands of years. The coast between Skegness and Mablethorpe, with its sprawling caravan parks – Golden Palm, Golden Sands, Country Meadows, Kingfisher – shouldn't be solid land at all, and perhaps that underlies their edgy brashness: precarity is implicit. Seventy thousand years ago, the area was under sea and the coastline ran along the edge of the Wolds, twenty miles away. Ten thousand years ago, you could have hiked from Skegness to Rotterdam over the undulating hills and valleys of the now-submerged Doggerland, with plenty of opportunities for good hunting and foraging in a fertile land; chances were that you would have found camps of others living in what archaeologists claim was a 'Mesolithic paradise'.

When sea levels began to rise in the North Sea at the rate of one to two metres a century, first the beaches disappeared, then the valleys filled up, leaving hills that formed an archipelago of small islands separated by fierce tidal currents and powerful tides. As the erosion accelerated, Doggerlanders moved west, to settle on England's seaboard in Yorkshire and Lincolnshire, no doubt bringing with them sagas and cosmologies prominently featuring floods as they tried to understand the catastrophe. Around 5,000 BC the last land bridge disappeared, and Britain was cut off from Continental Europe and the flow of people, ideas and innovations which had characterised thousands of years of prehistory.

In *The Making of the British Landscape*, Nicholas Crane outlines this historic moment and its far-reaching impacts: the tidal range of the Channel shrank dramatically as extensive mudflats and estuaries were submerged and crucial intertidal foraging land was lost. Plankton and the dependent food chain of fish and molluscs would have been severely disrupted. Britain was cut adrift from Europe as the North Sea's tidal rips, treacherous sandbanks and strong currents created by the now-submerged Doggerland became too dangerous to navigate in dugout canoes. Doggerland vanished out of human memory until the late nineteenth century, when fishermen began to bring up carved harpoons and animal bones. Archaeologists pieced together the ancient disaster, in what was described as an 'archaeology of absence': the study of what is not there, and of what is lost. The drowning of Doggerland led to increased territoriality – land became a precious resource as the sea rose, suggested archaeologists.

The Lincolnshire coast has a long history as provisional and prone to intermittent catastrophe. Is it fanciful to link that with the inclination evident in the area for clearly defined edges – the perfectly clipped hedges, trimmed lawns, picket-fenced pockets of garden and the painted kerbs of the caravan sites? In Daniel Defoe's great parable of Englishness, *Robinson Crusoe*, one of the first actions of the shipwrecked, stranded protagonist is to build a stout fence.

The rise in sea levels bequeathed a tussle between land and water on this soggy English eastern flank that has continued for millennia; the salt marshes were drained to create fertile fields, and dykes and pumps were erected to ensure that the water was kept in its place. As the currents keep shifting the shoreline, new land is even being created at Gibraltar Point, where a spit has begun to creep out tentatively into the Wash over the last twenty years. In another twenty, it may be possible to walk out on firm sand past the basking seals and gleaming mud and look down the Wash coastline, with its ditches, sluice gates and mechanical diggers shifting the ubiquitous mud. The coastline here has been the product of engineering – of steel and concrete and huge machinery to drain, gouge out, fill in and build up.

Meanwhile, another process of heavy construction has seeded the North Sea horizon with *fields* and *farms* of wind turbines, the nomenclature only adding to the muddled boundary between solid land and water. Once, holidaymakers on the beach in Skegness and Mablethorpe, sitting in their deckchairs looking east, saw nothing but sea between them and the Netherlands and Belgium. It was an unbroken horizon of choppy brown water, with its associations of space, wildness and nothing. Now the horizon is frosted with glinting silver pins of wind turbines, and their hypnotic slow whirling. In 2008 the then biggest wind farm in the world was built in view of the shore, and the repetitive horizontals of this coast – flat field, beach, sea, caravan roof – now contrast with the insistent verticality of hundreds of wind turbines. No longer an oceanic wilderness, the sea view has been industrialised, tons of steel and concrete sunk into the outer fringes of old Doggerland. Skegness visitors lucky enough to sunbathe (wind shelter essential) are essentially looking at a power station. There are plenty more to come. In 2020 the government announced a massive 355 per cent increase in capacity over the next ten years; that entails completing a turbine every day for the next ten years, yet experts on climate adaptation want even that ambitious goal doubled by 2050. The UK's seas are amongst the best locations for wind power in the world, and the best in Europe; being part of an Atlantic archipelago reaps big dividends. The wilderness that the sea represents is now forested with metal.

The Crown Estate owns most of the UK coastline out to a distance of twelve nautical miles, and during the last years of Queen Elizabeth II's reign, a fortune was made from selling licences for turbines which will benefit her descendants for the rest of this century. Wind power was a winning formula – ensuring the lights stayed on, eased environmental anxiety and, to top it all, a much-loved nonagenarian got a cut – and it led to the quiet acceptance of the biggest industrial intervention in our natural environment in recent decades. Who could complain?

The one catch was the poor kittiwake. The skies over the southern reaches of the North Sea are as busy as the shipping lanes below. Flamborough Head is England's biggest seabird colony, and the Hornsea Three wind farm is only seventy-five miles away. Research seems to indicate that the kittiwakes are disorientated by the thickets of metal, and they ineffectively slalom between the steel blades and poles to their deaths. Their numbers have declined catastrophically by 70 per cent since the 1980s, and the farms put them at further risk. Other birds such as guillemots and razorbills appear to be able to steer clear, but that requires detours which use precious energy needed for breeding. The Royal Society for the Protection of Birds is caught in a dilemma: the development of wind power is essential to tackle the climate emergency, which threatens millions of birds, but it presents a particular danger to the seabird population. A RSPB spokesman concluded bleakly, 'The North Sea is filling up with turbines.' Their numbers have been limited around the Greater Wash to protect its mudflats for the hundreds of thousands of lapwing, dunlin, knot, oystercatcher, bar-tailed godwit, Icelandic redshank, sanderling and plover, which rest to refuel before heading north to the Arctic breeding grounds. The Wash is effectively an avian global interchange, where 500,000 stop annually; it may be an unglamorous coastline, but it is one of the most important wetlands in the whole of the North Sea, from the tip of Norway to the shores of northern France.

The Lincolnshire coast veers between crowded – bank-holiday visitors, migrating birds, horizons full of turbines – and emptiness, with its huge skies and vast beaches. It may sound like a paradox, but it is a place of subtle extremes; as it slips from one to the other, one can catch the haunting sense of absence, of what has been lost, and the vanished lands of prehistory. And now, the more recent loss of the coastline's brief eighty-year heyday of buoyant vitality.

3

Dovercourt to Canvey Island

Essex

I am heading for the intricate, complex coastline of Essex, and a resort I've never heard of before, Dovercourt. According to Anthony Trollope, it was 'a not sufficiently well-known marine paradise', as he wrote in his 1876 novel, *The Prime Minister*. When I mentioned the town, people's reactions indicated that the first half of Trollope's phrase still rang true, but when I reached it, the other half was less evident. Dovercourt seems to provoke hyperbole; an online history site claimed that the bay of Dovercourt rivals the Bay of Naples. Neither a paradise nor Naples, the town has the sleepy, undistinguished ease provided by a pebbly beach, muddy water and concrete sea walls. One of its most endearing features is plentiful free parking (I was becoming a parking-app veteran, as every resort used a different one, all requiring large amounts of personal data).

The colour of the water was the familiar coffee. Unlike in Skegness, the weather was quiet and mild, with no intimidating North Sea breezes. As I swam, I had a good view of the cranes and container ships of Felixstowe to the north, while the nineteenth-century Martello tower of another resort, Walton-on-the-Naze, was visible to the south. The seaside resorts of this crowded coast jostle up against their rivals, and tolerantly accommodate as neighbours the bulky presence of major industrial infrastructure such as ports and refineries. The Essex resorts have never been associated with Romantic notions of natural wilderness or the picturesque. When *Country Life* magazine published a survey of the landscape qualities of the English counties in 2003, it harshly awarded Essex zero. This county cannot match the grandeur of Yorkshire or the spaciousness of Lincolnshire, but the inspiring work of writers, in particular Ken Worpole, has provoked a renewed interest in its coastline – second in length only to Cornwall. Its shifting, provisional and overlapping identities have struck a chord with contemporary readers and enthusiasts of psychogeography, Worpole argues, and pertinently asks, 'if landscape and national identity are uneasy familiars or surrogates of each other, it is worth asking what is Englishness today if its favoured topography is based on low horizons and the cold seas of its eastern approaches. Why does the zeitgeist now favour a lonelier, bleaker, more rebarbative sense of place?'

Essex has always had more than its fair share of those indeterminate spaces, often described as wasteland, where ruins are left to slowly disintegrate and rot, and which are used for landfill and scrap-metal yards. A coastline is always a work in progress, and parts of Essex's shore are raw, its future up for grabs. South Essex is the edge land of a global city, London, and its estuarine reaches have been used to manage the capital's voracious appetites – for logistics, ports, warehouses, oil refineries, power plants – and prodigious waste. Much of Essex lives with the gravitational pull of an indifferent London, akin to sleeping with an elephant. The city provides opportunities, and the morning commute on the A12 illustrates how it sucks in the workmen dedicated to keeping London's buildings in order – the electricians, plumbers, heating engineers, floor-layers, builders and decorators.

Dovercourt is no-nonsense these days, more forgotten suburb than marine paradise. Two old Victorian wrought-iron lighthouses feature in the images of the town, and there is little else to photograph. But, unlike many other resorts, Dovercourt doesn't seem burdened by its bustling past. Clues are discreetly tucked away – the grand statue of Queen Victoria built by public subscription, and the once-imposing Orwell Terrace with its handsome porticos and columns, which is now cheap bedsits. Dovercourt owed much to its proximity to the busy port of Harwich, serving as a genteel place to stay before or after the voyage to Germany; King Edward VII dined at the now-vanished Alexandra Hotel in 1905. By the interwar years, it boasted the modern attractions of a roller-skating rink, seven tennis courts, outdoor pool and then, in 1937, Warner's holiday camp opened, with a capacity for 1,500 guests, a ballroom with a 'satin-smooth oak dance floor', and such novelties as a wireless, radiogram, microphone and 'varied coloured lights radiated on to the dancers'. The camp's advertising celebrated the quantity of food: four meals a day in high season, it promised.

Within a year, it had been taken over by the Refugee Children's Movement, housing disorientated and homesick Jewish children fleeing Nazi Germany before they were dispersed around the country. Dovercourt, conveniently close to the ferry, was a staging point in a caravan of pain: a cruel asylum policy forced parents into an excruciating choice between sending their children away or risking their futures under Nazi persecution. Post-war, the camp was renovated, with novelties like discos for the teenagers, and 'two television rooms, one for the BBC and one for ITV'. Closed in 1990, no trace is left of the camp's fifty-three eventful years. Housing estates have steadily swallowed up Dovercourt's past: the camps, hotels, gardens, tennis courts, ballroom and skating rink have all vanished, leaving a neat, cheerful blandness.

Neighbouring Harwich (the towns adjoin each other) cannot bury its history so easily; in the 1720s, Daniel Defoe said it was a 'town of hurry and business, not much of gaiety and pleasure; yet the inhabitants seem warm in their nests and some of them are very rich'. That eighteenth-century wealth has left its traces in the fine houses squeezed into the narrow streets, and in the quaint pubs on street corners, while the infrastructure of the port – warehouses, offices and cranes – looms overhead and barges up against back gardens. Harwich is half heritage town, half port.

With some Covid restrictions still in place, it was eerily silent. In a restaurant window, the owner had put up laminated sheets with inspirational sayings: 'It's not the goal that matters but the journey, every step of the way.' When I paused to read them, the owner emerged to explain how he changed them regularly to help keep up morale. He had turned the restaurant into a takeaway for loyal locals, and was managing to keep busy he said, with stoical cheerfulness.

Samra Mayanja was five when she arrived in Dovercourt, the town that swallows history without leaving much trace. Her Ugandan-born mother had decided after her divorce that she wanted to bring up her two children where she could afford a house and a garden, and she continued to work in east London, in Newham, a commute of two hours. Separated from family and friends, Mayanja and her brother were the only children of minority ethnic heritage at the local school; the headteacher even saw fit to make an announcement of their arrival at an assembly, remembered Mayanja when we met.

'If Dovercourt was a box of photos, a lot of them seemed to me damaged at that time, their corners lost. I was determined to leave, my identity became all about leaving. I didn't care what racist shit I was dealing with, I took the view that ultimately I was beyond all that because I was going to leave. By the age of twelve, I was already sneaking on to the train to London.'

Mayanja had embarked on a project analogous to my own. Now a Leeds-based artist working across many disciplines, she has been making a series of podcasts on coastal towns entitled *Edging Home*. Drawing on her own experience of growing up on the coast, she explains in the introduction to her podcasts that, 'I want to talk about England. I want to talk about the permaculture of ideas around edges. [These are] places of great biodiversity. The edge of time, edge of history and edge of the nation. I am afraid. I want to talk about building bridges with people that I fear hate me in theory.' She concludes with one of her favourite sayings: 'To travel the world is scary but to knock on your neighbours' door is truly adventurous.'

As she was growing up, Mayanja was lucky to find support from a few teachers; one once handed her a Tesco carrier bag full of books written by black women, Maya Angelou and Toni Morrison among them. 'I think about her a lot,' Mayanja commented. She did well in her tough school and got a place at York University.

'The rain in Harwich is soft, light and gentle – not like Leeds – and the sun hits everything,' she remembers. 'I don't go back much. I did a residency in Barrow-in-Furness recently, and I find that on the coast, you have a strong picture of England and its relationship to the rest of the world. On the coast, there is potential to speak across many topics – the economic, social, emotional, and the weather; they are all in flux, they crash and loop around like the rhythms of the sea. I had the sense when I visited Bridlington that there was something gorgeous, but it had gone. No one knew how or where to find it. Barrow had an industrial history of shipbuilding and that gave them a pride, but you don't have that in Harwich. That's really significant. Harwich and Dovercourt are places where history doesn't pass, it just carries on hanging around.'

Just south of Dovercourt, the coast returns to its original muddy self. Hamford Water forms a deep indentation in the coast, as if someone has taken a large mouthful and not been able to chew it all; the inlet is scattered with islands such as Horsey Island, Garnham's Island and Skipper's Island. Bramble Island was used to test explosives until 1985, and the company who owned it exported hazardous chemicals to the US until 2009. Between the islands a pattern of creeks and rivulets – one is known as the Twizzle – expand and contract with the ebb and flow of the tide. On a map, they look like the cracking of fine glaze on an ancient Chinese vase.

Early one morning, I arrived at Hamford, intent on reaching the biggest island, Horsey, by bike. It was a distraction from my real goal for the day, visiting the string of resorts further south, from Walton-on-the-Naze down to Jaywick, but I was intrigued by this pocket of wilderness and the muddy coastline out of which all these resorts emerged. A narrow lane to the shore ran past an aggregates yard, where lorries were being filled with gravel. Next door was a course for dirt bikes, with piled-up rubber tyres. But the way was barred by a padlocked gate, a common feature of Essex's coastline; the place was bristling with its own anxious territorialism. I lifted my bike over the gate, and before long I was at the edge of the water, beside a derelict

Second World War gun emplacement and a rusting, padlocked container. On one side were PRIVATE, KEEP OUT signs, while on the other a path along the top of the dyke skirted the mud and marsh between verges brimming with wildflowers. A damp mist blurred the outlines of water and land. Ahead, a ramp led down to the water, and the causeway to Horsey Island still a few inches deep in the silvered water of the receding tide. The island's dim outline was just visible through binoculars. The causeway is only passable for four hours at low tide.

As I pedalled across the sea, water was ebbing on both sides in skeins of silvery rivulets, the mud gleaming. Halfway across, I got off and stood in the seawater to marvel at my immersion in this range of greys – a vista of cloud reflections and mud – with few points to fix myself: I was the only thing that was vertical. Only a thin line of dark grey indicated Horsey Island's wooden jetty and reassured me that I wasn't in danger of cycling into the North Sea. As I came closer, I could see land, scattered with the debris of boats, while cormorants sat on the seaweed-draped timbers, their wings hung out to dry. Despite a KEEP OUT sign, I set off along the dyke which surrounded the island; a Range Rover appeared almost immediately, and headed straight for me, driving over the meadows at a good pace. The driver was insistent that I leave immediately: I was disturbing nesting birds. He had the self-assurance of one used to turning people off his land, and I later reflected that someone who chose to live on an island which is inaccessible most of the time is unlikely to be well-disposed to passers-by. The driver's defence of his escapist, lonely island provoked more fascination than irritation; even in the densely populated south-east of England, it was possible to find a personal fiefdom on this muddy coastline. I retreated back along the causeway, watching oystercatchers fastidiously

pull their webbed feet out of the sticky mud with each step, using their long beaks to forage for food. As the tide retreated, the mud quietly bubbled and hissed.

Once back in the car, a ten-minute drive took me to the small seaside resort of Walton-on-the-Naze. I'd swapped the mud of Hamford Water for a busy car park, and a parent was sternly insisting, 'No, you are not having an ice cream.' Essex's versions of itself are sharply contrasting – the muddy marsh, the neat bungalows, the shabby pier, the brash casino – and are dovetailed to fit alongside each other. As I unloaded my filthy bike, my boots claggy with the marine mud, trousers and raincoat spattered, a young couple emerged from a neighbouring vehicle. Despite the chilly sea mist, the woman was dressed in a scarlet silk minidress, the sleeves falling off her tanned shoulders, gold chain gleaming around her neck. I gawped: there seemed no better example of the strangeness of Essex, its sheer implausibility. Was she not cold?

The cliffs on the Naze headland are being eroded at the rate of two metres a year, which means that the row of immaculate bungalows with their magnificent sea views probably have about twenty years before their manicured front lawns and pots of geraniums are in danger of toppling into the North Sea. They will be joining the remains of earlier versions of Walton-on-the-Naze, which lie under the sea out to a range of nine miles. A prebendal stall in St Paul's Cathedral in London is dedicated to the town, and carved with '*Consumpta per mare*' – consumed by the sea. A bungalow owner was carefully mowing the grass around a rose bush on the edge of a cliff which was soft and crumbly like fudge; trimmed hedges, lawns and flower beds have become a form of impotent protest against fierce winter gales and an irritable North Sea.

Sea walls run continuously from the Naze, past Walton, Frinton, Clacton and Jaywick. The resorts sit close together, but each has a distinct character, clientele and class profile: a builder friend, born and bred in the East End, was well versed in the gradations of status evident on that coastline, where many of his family have settled; he mentioned a friend who had hit a rough patch and 'ended up in Jaywick, even though she had been born in Walton'. The profile of the towns is shaped by the great inequalities of London's wealth, which ripple out to nearby coasts.

Walton's concrete promenade is lined for much of its length by beach huts, and as I cycled south past the shabby amusement arcades towards Frinton-on-Sea, their furnishings became more elaborate. One family had decorated the entire hut with retro floral Cath Kidston prints: tablecloth, deckchairs, teapot and mugs. By the green sward of Frinton's golf course, my bike came to a halt with a puncture: the ride on Horsey Island's pitted causeway had caught up with me. I locked the bike up and set off to investigate the town while waiting for backup and a puncture repair kit.

My grandmother always holidayed in Frinton, and every year her postcard arrived, to sit propped on the mantlepiece for a few days. Memory is stuffed with such inconsequential detail and, out of it, we accumulate a geography for our family history. As a child, I was puzzled by those postcards of the flower beds and lawns of Frinton in the 1970s. On the back, there were always a few perfunctory comments about the weather in my grandmother's round, old-fashioned writing. My relationship with the seaside was about camping: Runswick Bay's streams running down through thickets of dog rose, bramble and bracken on to the boulders of the beach; only the sea to wash in and a delightful fortnight of grubbiness, sun (if we were lucky), mud and sand. My grandmother's holiday was something else entirely – but what?

From the start of its development in the late nineteenth century, Frinton was intent on keeping what it perceived as the riff-raff at bay. Draconian by-laws banned pubs, cyclists, piers and all forms of commercial activity on the seafront, even cafes and ice-cream parlours, and definitely amusement arcades. The names of the streets make the town's intentions plain: Oxford, Cambridge, Eton, Queen's Road and Kings Home. Frinton was designed to appeal to the anxiously class conscious, and proudly boasted that Edward VII frequented the golf course, and Winston Churchill had rented a house. On the broad, tree-lined avenues, roses tumbled around the windows of the large Edwardian houses, and well-maintained gardens brimmed with shrubs. Positioned between Walton and Clacton, it was determined to have nothing to do with either of its neighbours. Famously, railway gates separated Frinton from Clacton, and when they were finally dismantled in 2009, in the face of dogged protest, it had to be done at night by Network Rail.

I imagined my grandmother walking briskly along the sea-front in her blue crimplene suit, a scarf over her white waved hair, wearing her double string of pearls. She holidayed alone. When she sent those postcards, she had been a widow for eight years, dependent on her daughter and son-in-law. Frinton was her only chance of a holiday, my aunt explained to me. My grandmother's sister had inherited the family fortune and bought a flat in Frinton, which she lent to my grandmother for a fortnight once a year. It came with a beach hut. That fortnight was the highlight of my grandmother's year. But such were her straightened circumstances, she had to plan her meals with care. One cheap staple was bacon shank, my aunt remembers.

I inherited my father's irritation with his mother – perhaps unfairly. Her start in life was financially precarious, as a child of a single mother, and she ended up living her last three decades dependent again on family support. Her persistent preoccupation was the English class system and, in particular, her family's upward mobility. But as I looked at Frinton's quiet, grey sea, I softened. What chance did women of her time have? Behind her snobbery was a defensive Irish migrant story of ambition and hidden poverty. I hoped Frinton brought Bridie pleasure, soothing her social insecurities with its reassuring gentility.

By now it was raining softly and I hadn't yet had a swim. I peeled off layers, and, shivering in the cold, I raced across the sand. A passer-by wrapped in a thick overcoat looked horrified. 'You're brave.' Back in the car, I wriggled out of a wet, sandy swimsuit as the rain thrummed a steady beat on the roof.

Another lockdown is easing and I'm in Clacton on a chilly April day. As I arrive at the beach, the sun breaks through the clouds and I sit down on the sand, revelling in its shy spring warmth, watching the high tide recede and the wind turbines on the horizon lazily turn in the still air. Next to me, a small boy is playing with his grandfather; his mother and grandmother are chatting on the bench behind. Every generation relishing the ease and calm of the quiet waves and fresh air. On the promenade, the benches are full as people enjoy the simple pleasures of meeting and talking as they sip coffee. The seaside's appeal spans every age group; later, as I leave the town at 5 p.m., teenagers emerge on to the now empty promenade, and a gang of boys leap over the railings to dive down behind the lavatory block for some mischief.

By late morning the next day, the fryers are already at work for lunch, and the smell of hot oil hangs heavy. From a stall selling buckets and spades comes Adele's 'Chasing Pavements'. The combination of salty fresh air, dated pop-music favourites and sunshine seems to sum up the English seaside: familiar, nostalgic, and deeply comforting.

Lin and her father-in-law, Fred, are sitting out in deckchairs in front of their beach hut, and are happy to be interviewed, so I sit on the wall beside them. Lin has a quick, warm smile. She is the mother of five children, aged six to eighteen, several of whom she has brought to the beach with their friends, and our conversation is punctuated by their comings and goings, as she hands out snacks and drinks. They have rented their hut, known as Damarkand, for £1,000 a year.

'We're local, we're from Holland,' said Lin. For a moment I'm confused, but then realise she is talking of the village, Holland-on-Sea, not the country. 'We spend a lot of time on the beach, so we thought, why not get a hut – my father-in-law plans to come down here every day.' Picnic things lie scattered on a table. A passing couple joke that they will be back for a cup of tea, and could Fred put the kettle on? 'People stop by and chat – like you,' she adds.

Lin moved to the Essex coast from East Ham in London when she was eleven, and, to start with, hated it. 'Couldn't wait to get back to London at first, but I've never left. I love the space and the quality of life here. The kids can play outside. I would never go back.'

Fred moved here thirty-four years ago, having grown up in Hackney Wick, and he warms to the subject of the East End, describing the nights out he used to have at the pie-and-mash shop in Dalston, and rattling off the names of his favourite pubs. 'We're all Londoners round here. We're *proper* people.' I ask him what he means by the adjective, and, by way of reply, he explains that, 'East Londoners moved to Clacton and Southend, and south Londoners moved to Brighton.

'I was thirty-eight when I came here,' he continues. 'We had a bungalow as a holiday home, and I bought a fruit-and-veg shop – worst decision of my life. It took me a year to build it up, and a year to sell it – at one point I was earning 50p an hour. I'd been a carpenter in London, so I went back to that and got a job doing conservatories – that worked out well, and I set up with my son. We moved into double glazing and the business lasted twenty-five years; my son took it over, but it fizzled out.'

Lin is emphatic: 'London is not what it was. Even in Ilford there used to be nice areas, but it's all changed. Or perhaps it's me that's changed; my views have changed. It's the crime that bothers me. I didn't want to bring my children up around that. There's a good community feeling round here.'

When I press her to explain what she means by 'changed', she can't or won't explain; her father-in-law agrees that the crime in London is terrible. I am wondering if 'change' and 'crime' are coded ways of talking about immigration – Clacton's seafront is largely white – but I sense a slight unease. Clacton's political history has brought more than its fair share of visiting journalists eager for ways to judge the town. I let Lin take the conversation in the direction she wants. She declares herself very happy with the local schools and training opportunities, explaining how one of her sons is at college studying catering and working in a chip shop.

'My kids can do anything here if they set their minds to it. I know housing is expensive, but it's up to them. If they want to get a job in London, they can commute – it's an hour and a half on the train. I hope they stay round here, it would be lovely. My husband has always lived next to his parents, and I live close to mine.'

Her father-in-law has disappeared into the hut, and he re-emerges holding a substantial hardback book, *My East End*, to show me. 'I remember the chopped eels on trays – my mum and dad loved them. I'm proud to have been born in Hackney Wick. My mother was Irish and I was born in 1946 – we behaved ourselves, and we were safe. I wouldn't feel safe there now. I used to drink eight pints an evening and still drive –' he laughs at the memory. 'I'd be robbed blind if I did that now.'

I'm puzzled again at the reference to safety – especially after we all agree that his drinking and driving sounds very dangerous – but the conversation moves on to holidays.

'Of course we go to the seaside,' laughs Lin. 'Usually Butlin's at Minehead or Skegness. Bognor is too expensive. We went to Tenerife two years ago.'

Fred is back to his memories: 'My childhood holidays were at Valley Farm, here in Clacton. Four hundred caravans. We enjoyed it so much that we thought we might as well move here.'

'I don't like the drug crime in the town,' admits Lin. 'There's more and more, because the access is so easy from London. And it's a shame that the shops in the town centre are closing down, but people buy online – we're as much to blame as anyone.'

Next door to Clacton is Jaywick, which is the poorest place in the UK, repeatedly topping tables of deprivation over the last decade, each time garnering headlines and attracting camera crews for news items and documentaries. Lin used to work in the doctor's surgery there. 'I know it's very poor, but there is an incredible community spirit. When you have very little, you help each other. One lady used to cook big dinners for anyone who needed food. Jaywick is misunderstood. When people are in a rut, it's very difficult to get out. I saw how if you are on a very limited budget, you can't afford to go to the chemist for something; I could see how much people struggled. People didn't know how to cook or how to use leftovers. My nan would cook a roast on a Sunday, and then use the leftovers for the rest of the week, but these people don't seem to have learnt that kind of thing – I don't know how it broke down.' For a moment, the warm-hearted Lin looks confused, puzzled by lives which have not had her structure of close family.

I tell Lin and Fred that I'm off to swim and they look aghast, joking that I'll come back shivering with cold and blue lips. The water is indeed icy, but once I'm back on the beach and rubbing dry, I tingle with the familiar rush of vitality. Hot fish and chips bring warmth back into my numb fingers. I lick the grease off with satisfaction, pondering this east London diaspora with its strong sense of family, pride and tough dignity. Unwittingly, I've benefited from their displacement, having lived for several decades in the neighbourhood where Fred grew up. Many, like Fred, drifted out of London to the coastal edge – in search of space, homes with gardens, dreams, a change or a staging post, where they got stuck.

Back on the promenade, I stop another visitor; Elaine is

reluctant to talk, her face worn and anxious with a strained sadness, but when I ask her about Clacton, the lines ease and warmth is evident in her smile.

'This is what I need, it's lovely,' she says, looking past me at the calm, blue-brown North Sea. 'I want to move here.' She adds shyly, 'My dream is to live in a seafront flat.' Her eyes gleam at the idea. She says she is a care worker, and is now living in Suffolk, but it is Clacton which 'feels like home', and she comes at least five times a year. 'It's always got to be Clacton – I've childhood memories of holidays here.'

I ask if her move to Clacton will be retirement, estimating that she is well into her sixties.

'Oh, no, but I'll find work here easily. There is plenty of care work. I love Clacton, even though people say it's crap and ask me why I want to move here. It is dated, I know, but I love it.'

In the town centre, lined with charity shops, a group on their mobility scooters have gathered to chat, while at another bench a cluster of men argue vociferously, their voices slurred with alcohol. Kay is walking briskly down the pavement with a small dog on a lead; she is smartly dressed, her pink lipstick matching her jumper, and wearing knee-high patent-leather boots. She says she can only spare a moment to talk. She came from Bermondsey, London, twenty years ago, but admits she doesn't know why she and her husband came nor why they have stayed so long. Now, her eighty-two-year-old husband doesn't want to move. She doesn't have much time for Clacton, she adds bluntly.

'I can't think of things I like here, but I'm of the opinion that you make the best of your life. It's not about where you are, it's about who you are. I run on the seafront, I don't allow bad thoughts, I just tell them to get out of my mind.'

Kay launches into her life story. 'My father insisted that my sister was given away – he was an abusive man. There was only eighteen months between us. She ended up in Australia. I'm a descendant of Mary Queen of Scots – look at my red hair – I've had two businesses employing sixty people, and six children. I'm seventy-five and still working.'

I express admiration and she relaxes, pulling on the dog's lead so that it lies down, as if to listen to her story.

'My family was poor, I went to work at fifteen. After I had children, I set up a cleaning business by going from house to house, with the buggy, leafletting, and then I set up two video shops. I work as a tailor now – I learnt it as a child at home. I would go to jumble sales and buy clothes to unpick and resew. I've always been a striver. I'm a powerful woman.

'After we came here, I went back to college and did computers. I speak four languages – German, French, Spanish, English. I do a lot of voluntary work on mental health – my eldest son committed suicide, but I never say "poor me". I can see when people need help, I've got that intuition.

'I don't volunteer for any organisation – they would never have me. I have my methods.' She offered an example of a neighbour who was begging for help; his wife wouldn't get out of bed after their son's suicide. 'He said to me, "Kay, I know you've been through this." So I went up to her room and I told her, "You have two choices; you can jump off the roof with me now, or you can come for a walk: which is it to be?" No organisation or charity would let me say that kind of thing. But it worked – I see her out and about now.

'I come from a generation where you got on with it. We didn't realise we were poor. You just went out there and made things better for yourself. My son said to me, "We can't all be like you," and it's true, people don't always know my strength. It goes back to Mary Queen of Scots.

'Money doesn't make you happy. Clients come from London

for my tailoring; one of them was looking down his nose at me. I know the type, and I said, "I don't need your arrogance. You're no different from anyone." He was shocked, he had never been spoken to like that. I told him. "I don't want your money, you can't buy me." He left without paying. He had the problem, not me. I'm an independent woman. I've got my own values.'

Clacton has a big reputation, and provokes both affection and loathing. In 2014 its Conservative MP, Douglas Carswell, defected to the Eurosceptic UK Independence Party and provoked a by-election. Clacton stood by their right-wing libertarian MP, and he became the only UKIP MP, adding to the pressure on the then Conservative prime minister, David Cameron, to allow the referendum on British membership of the European

Union. Clacton proved to be pivotal in the internal Tory Party struggle over Europe, as Carswell acknowledged in an interview in 2020; he claimed that 'the revolution started in Clacton', and that the town 'had changed the course of the country's history'; in the 2014 by-election, 'Clacton folk forced politicians to agree to the referendum.'

Carswell's election victory prompted a coruscating column in *The Times* by the Conservative commentator Matthew Parris. Under the headline 'Tories should turn their backs on Clacton', Parris wrote that 'only in Asmara after Eritrea's bloody war have I encountered a greater proportion of citizens on crutches or in wheelchairs ... this is Britain on crutches. This is tracksuit-and-trainers Britain, tattoo-parlour Britain, all our yesterdays' Britain ... I am not arguing that we should be careless of the needs of struggling people and places such as Clacton. But I am arguing – if I am honest – that we should be careless of their opinions.' With the hindsight derived from the Brexit victory of 2016, Parris's view is damning evidence of why millions of voters rejected what they regarded as the Remain liberal metropolitan establishment. Mutual contempt had split Britain into passionately opposed camps. Clacton may have appeared to Parris as 'all our yesterdays', but it turned out to be all our tomorrows, triggering a momentous upheaval in British politics and its relationship to the world.

Carswell was dismissed as an irritant by Conservative commentators in 2014, but after Boris Johnson became prime minister, Carswell returned to the Conservative Party, and in November 2020 he was appointed a trade advisor. Despite being back in political favour, he said he had lost hope of the British people ever voting for full-blooded libertarianism; in January 2021 he became head of a small think tank, the Mississippi Center for Public Policy, in Jackson, in America's Deep South. Announcing his quixotic career decision, he declared that the

US was 'under attack' from a 'radical New Left'. Having pulled off an establishment coup in Clacton, a UK backwater, the forty-nine-year-old Carswell appeared intent on repeating the trick in another such place in the US, well away from the metropolitan hubs he criticises, of New York, Chicago and California.

Clacton and the wider district of Tendring differ from neighbouring districts in two respects. They are disproportionately old and white: 95 per cent of the population is white British, well above the national average of 79 per cent; nearly 20 per cent of households are a single person aged over sixty-five. It also has a very high dependency rate, of one working-age person to eight over sixty-fives. The demography has played a key role in its politics. Ken Worpole suggested in a conversation with me that the key to the stretch of Essex coastline from Clacton to Canvey Island on the Thames Estuary is its 'continuing, symbiotic relationship with London's historic East End'. Loyalty to family and community are highly prized, Worpole added, and 'patriotism now fills the gap left by the decline of the spirit of working-class non-conformism as the loyalty to a higher cause'. These resorts have been shaped by 'the male working-class culture of the East End: the informal economy, street markets, ducking and diving, bunking off to sea if things got rough, being your own boss,' he continued. 'If the northern working-class hero of the 1960s was factory machinist Albert Finney in *Saturday Night and Sunday Morning*, then the London hero was Alfie, a picaresque, artful dodger. I just sum it up as the feeling that "nobody has a right to tell me how to live my life".' One of the biggest events of the year is Clacton's annual air show, in which Lancasters, Spitfires and Harrier Jump Jets perform for thousands of spectators. This is a town proud of British military history, and plenty of flags were in evidence – both Union Jacks and flags of St George – flying from masts in private gardens.

In London the builder who regularly does work on houses in my street in Hackney tells me he's already booked into Clacton for the August bank-holiday weekend. He was not looking forward to it, and laughed as he explained: 'All my wife's extended family go to Clacton, and we rent adjacent caravans. All they talk about is houses and money. Since my wife and I are the last ones left in the East End, they think we're the losers. The rest sold their London flats after they had made some money and moved out to Essex, and have been buying and doing up houses ever since. But they still go to Clacton for the bank holiday – it's a family tradition, no matter how much money you've made.'

At the beginning of the twentieth century, faith in the benefits of fresh air and space for every form of social disadvantage made this part of Essex a therapeutic dormitory. Examples included Temperance Hotel on Osea Island, founded by Frederick Charrington, a brewing magnate with an uneasy conscience, and later adapted as a sanctuary for those with addiction problems; and the Girls' Bungalow, established in 1909 by the trade unionist Clara James as a holiday home for working girls from London. Crippleage in Clacton was established to provide a home and training for orphaned handicapped girls from London, and residents were taught to make silk flowers. Essex became a form of political laboratory, Worpole maintains, testing out new ideas of community living, many of which originated in the East End's maelstrom of trade-union activism and left-wing politics. Camps were set up – chalets or tents – to offer affordable holidays which could also inculcate the socialist camaraderie needed for effective political change. Another experiment was the plotlands: small parcels of land were sold as leaseholds, where purchasers could build their own shelter. At the peak of their popularity, they were evident in many parts of the UK, and particularly along the east coast. The chalet where

I stayed on my childhood holidays in Runswick was part of this popular movement for cheap holidays in the 1920s and 1930s.

The small resort of Jaywick started life as this kind of plot-land in 1928, when an entrepreneur, Frank Stedman, reclaimed low-lying marshland just south of Clacton and sold plots to the better-off residents of Stepney, Poplar and West Ham. Daily coaches ran in summer from London's East End to Jaywick. Stedman named the roads of his development after the brands of car he loved – Buick, Daimler, Triumph – and laid out the town in a grid pattern representing a Daimler radiator. Purchasers were left to construct their own chalets, DIY-style. The problems began early, after Stedman was slow to bring in services such as mains water, sewage and electricity. The tracks became rutted, the buildings were mostly wooden huts built on the cheap. His promise of tennis courts, a pool, a licensed club and a lake failed to materialise. Clacton council became alarmed that some families were using the huts not just for holidays but as their primary residence, and argued that the site was inappropriate for housing because of its vulnerability to flooding. By 1936 Stedman, the council and the residents were at loggerheads, and the issue went to court, in legal cases which dragged on for years.

From the beginning, Jaywick divided opinion. Some argued that Stedman was a philanthropist trying to make coastal England affordable for the working class; George Lansbury, the Labour MP for Poplar, and briefly Labour Party leader, saw Jaywick as the beginning of something much greater: 'I just long to see a start made on this job of reclaiming, and recreating, rural England.' But other contemporaries were horrified, including the influential architect of post-war planning, Patrick Abercrombie, who lamented that in plotlands like Jaywick, 'innumerable wooden shanties have sprung up [and are] artistically deplorable [so that] whole fields have become so packed with them

that they are extremely unsanitary ... the preserver of rural
amenities cannot allow any sort of old junk cabin to deform the
choicest spots'. In 1938 another critic went even further, pouring
contempt on the rapidly expanding coastal developments:

> All is changed today in the English sea-villages. Nothing
> but a dictatorship will save the English coast in our time ...
> When the millennium arrives, when battle ships are turned
> into floating world cruising universities, perhaps their guns,
> as a last act before being spiked, will be allowed to blow to
> dust the hideous, continuous and disfiguring chain of hotels,
> houses and huts which by then will have completely encircled
> these islands.

Many coastal plotlands were in fact destroyed as a defensive
measure in the Second World War, and in 1947 a legislative ban
on their development came into force.

But Jaywick survived, and housing shortages in the immedi-
ate aftermath of the war led to more people living in the town
permanently. There were floods, as the council feared, in 1948
and 1949. Residents clubbed together to build a new sea wall in
1951, but it could not protect the town from the massive North
Sea flood of 1953, and Jaywick suffered one of the worst casualty
rates in the UK, with the loss of thirty-five lives and extensive
damage. A long struggle followed, in which the council sought
to demolish part of the town, then refused to put in infrastructure and imposed restrictions on further building. More floods
followed, until in 1988 new defences finally brought a degree of
safety, but at the cost of any sea view. The beautiful blond sands
of Jaywick disappeared behind a concrete façade.

The town still bears the marks of its origins a century ago:
some streets look temporary, the small houses more like summer

houses than permanent homes, squeezed close to their neighbours on small plots. Many are in a poor state of repair, and on
some 'streets' the front gardens are heaped up with rubbish on
either side of the pitted dirt tracks. Better-quality housing has
been built on the outskirts of the town more recently, but the
core is still recognisably a 1920s plotland. In the last decades of
the twentieth century, the attractions closed down – including
Butlin's in 1983; the visitors dried up and the cost of housing fell,
attracting absentee landlords who rented property out cheaply
on short tenancies, confident of securing a regular revenue from
tenants on housing benefit. That brought in newcomers – many
from London – whose lives had hit the bottom and were looking
for a new start; perhaps they had fond memories of childhood
holidays, and were chasing the dream of turning things around.
But they brought with them their demons, such as addiction,
mental ill health, disability and chronic illness. While some of
the chalets are lovingly cared for, others are not maintained by
absentee landlords, and tenants struggle to get the help or services they need to get their lives back on track.

Perhaps one of the most bizarre chapters of Jaywick's recent
history was when an image of the town was used in the campaign
of a Republican candidate in the 2018 US mid-term elections.
Over the photo of a rubbish-strewn street, voters were urged to
keep America on track and avoid foreclosures. The mistake was
gleefully emblazoned across tabloid front pages in the UK, but
for Jaywick it was a further bitter humiliation. It had effectively
become a global cliché of what poverty looked like in a developed Western nation – a symbol of economic and social failure
that had become relevant and transferable anywhere.

The most detailed government survey, the four-yearly Index
of Multiple Deprivation, compiled according to a range of
measures such as income, crime, education levels and disability,

named Jaywick as the most deprived place in England for the third time in 2019. It also showed that the wider district of Tendring, covering parts of West Clacton which have a similar profile, is becoming poorer, while many other parts of Essex are getting richer. The 2021 Chief Medical Officer's report into seaside towns featured the area as a case study because of its deep-seated problems; Clacton has the second-highest mental-health need in the country (only Chesterfield is worse), and a high suicide rate. Rates of self-harm are 50 per cent higher than the national average, and the suicide rate for men is 66 per cent higher than the national average and 89 per cent higher for women; the latter is striking, as women are generally much less likely to commit suicide than men.

Given that Tendring includes prosperous areas such as Frinton, the Chief Medical Officer's report added that only when the data is collected at a more granular level will the 'severity and rate of decline' in some places be evident. If only Clacton wards are considered, the level of childhood deprivation 'exceeds all comparators', it commented. In common with many seaside towns, the availability of cheap, poor-quality food sets up a pattern of ill health early. A quarter of pupils in reception are classed as overweight and amongst those over eighteen it is nearly 70 per cent. It has some of the worst educational achievement in the country: only 27 per cent achieved grades 9–5 at GCSE in English and Maths in 2018, before the pandemic's disruption of schools. Low aspirations and the lack of good job opportunities in the area are blamed for low attainment; unemployment was 50 per cent before the pandemic, and given how much employment was in sectors affected by lockdowns, Covid further exacerbated the plight of these entrenched pockets of poverty.

The media coverage has been damning; a *Daily Mail* headline once described Jaywick as 'Misery by the Sea', and the town has attracted more than its fair share of documentaries with titles such as *Benefits by the Sea*. One of the more thoughtful films, *Jaywick Escapes*, offered an affectionate portrayal of a tightly knit community, with bingo clubs, exercise classes and a busy summer fair, but it acknowledged Jaywick's influx of troubled incomers fleeing London in search of cheap housing and a new start: one interviewee was escaping the entanglement of drug use; for another it was the grief of a lost partner. Tragedy, much of it originating in London, is not hard to find in Jaywick. But a defiant pride was also evident on my visit; Union Jacks and St George flags were flying on some of the more prosperous properties, which had built decorative balconies overlooking the beach. The characterful pub was named Never Say Die, while one house was wreathed in Gay Pride rainbow flags and slogans asking people to pick up their rubbish. Some homes were looked after with great care by residents who have fought doggedly for the town's survival.

Now, in a revisionist reading of Jaywick's long history of conflict, reluctant council support and planners' contempt, one critic argues that 'given the right supports and time to evolve, shanties could have provided mass-scale housing . . . the process of self-build, improvisational design and adventure that was involved, deeply changed the people involved'. This was a rare and valuable experience of working-class people taking control of the construction process. Another critic acidly concludes that the 1960s high-rise developments were a much worse failure than Jaywick's self-build plotlands.

Before arriving at Essex's biggest resort, Southend, I wanted to see more of the county's muddy marshes, so I chose a circuitous route via a brand-new island. Wallasea has been created out of three million tonnes of earth dug from under the streets of London as part of the construction of the Crossrail tunnel and shafts. It is now a bird reserve a few miles north of the city of Southend. There could be no better metaphor for how London spills over, quite literally, on to this Essex coast. The intertidal area of 115 hectares of salt marsh, mudflats, saline lagoons and grazing marsh looks a little raw, with swathes of recently turned earth and fresh gravel. It has been designed to create every kind of habitat and water depth required by a range of migratory birds, and, more disturbingly, to accommodate the growing surge tides which might threaten flooding in London due to climate change. Four centuries ago the Essex coast was a haven for wildlife, but a combination of agriculture, coastal erosion and rising sea levels have reduced intertidal salt marsh by 91 per cent.

At 8.30 a.m. few people were about; it was just marshes, hundreds of avocets and terns and, in the distance, the crowded skyline of Southend's tower blocks. It felt like I was spying on the resort as I studied its profile through binoculars across the ruined hulks of old boats rotting in the muddy creeks. Larks swooped overhead in the cloudless blue sky, and their song filled the air. A couple of egrets shone brilliant white against the green grasses and blue sky.

The sea is ten miles from Wallasea, reached through a maze of creeks squeezing between the islands of Potton, Rushley and Foulness. The tides creep deep inland and when they recede, they leave a rim of dried grass and debris to show the high-water mark. Looking out over waving grasses, a sail was visible, moving between the fields on the tidal river. Water was everywhere in an uneasy intricacy in this sodden 'land', and the wide banks of mud exposed at low tide shone as if polished in the sun. In one creek a rusted boat had come to rest, its windows and doors gaping to the elements.

The Wallasea reserve is another chapter in Essex's history of absorption and accommodation of the regular influx of water, birds and people – the day trippers, retirees, London commuters and East Enders in search of a better life. The birds haven't hesitated, already at home in Wallasea's lagoons, no doubt appreciating the rare quiet on this busy coastline, crowded with human presence. Here is Essex's muddy, wayward coastline, before the invention of concrete. For centuries, efforts have been made to corral the water, to hold the line between land and sea, with embankments of timber and dykes. Their rotting remains are evident in many places. It was reinforced concrete which appeared finally to offer control, predictability and stability on this restless coast. As the coast turns into the Thames estuary, sea walls keep the river in check, thus creating the seven-mile seafront of Southend.

Nothing in Southend betrayed its muddy background. In the bright sunshine, the clear water sparkled invitingly on the south-facing beach, which looks out over the estuary to the chimney stacks of the north Kent oil refineries, where a heat haze blurred the outline of infrastructure at Grain Island's port. Southend's pleasures are not about the picturesque, nor even about expansive North Sea horizons; it's an estuary resort, umbilically linked to London, its beach offering a ringside seat to watch the freight ships heading upstream to the new London Gateway port. Southend-on-Sea has a distinguished history of entertaining Londoners for more than two centuries. In the eighteenth century, crescents and terraces were built to accommodate the visitors, including benighted Princess Caroline, the scorned wife of the unfaithful Prince Regent. The building of the railway in 1856 from London's Liverpool Street launched its heyday as one of the capital's most popular resorts, particularly for visitors from the East End. The Palace Hotel (formerly the Metropole) was once the only five-star hotel on the south-east coast, with 200 bedrooms and a magnificent ballroom. Like many other resorts, Southend fell on hard times in the last decades of the twentieth century, but rising London house prices have helped its reinvention as a successful commuter town. A university, art gallery and airport have brought it a new confidence. The prosperity of the town is evident in the tree-lined avenues of detached houses, and the large seafront mansions with glass balconies and chrome railings by Thorpe Bay. In 2018 the *Spectator* admitted, to its surprise, that despite being 'slightly scruffy', Southend was a fine place, and was puzzled that 'it has been a running joke for as long as I can remember. Why do people take the piss out of Southend?' before praising it for its lack of airs and graces.

On the seafront lined with palm trees, Southend had a breezy, unabashed spirit. I watched a tanned young woman walking along the promenade, her hair blowing in the breeze, in a bright summer dress and high-heeled sandals. She was enjoying herself. In a car park on the seafront two women had laid out their towels on the gravel to sunbathe using the car doors as windbreaks. A couple on a quiet stretch of beach were wrapped in a passionate embrace, the bikini-clad young woman lying on top of her boyfriend; they seemed close to orgasm and everyone averted their eyes, too polite to take any notice. Later she lay with her head propped up against her partner's tattooed chest, one elegant tanned hand languidly draped over him, her nails a deep red. Few were in the water, but, judging by the spectacular tans, Covid-furloughed Britain had used the strange, hot spring of 2020 to sunbathe. Brown flesh was bared to expose plunging cleavages, bare torsos, bulging midriffs, bellies and plump thighs. Southend, like most other English beaches, offers plenty of evidence of the nation's obesity crisis (in 2019, 28 per cent of the English population was obese, and 36 per cent was overweight) and its causes: the lavish supplies of the beach diet of chips and ice cream (slathered in cream, chocolate and jam sauces).

In his account of the English seaside, *Wish You Were Here*, Travis Elborough argues that 'like the NHS, the beach is one of the great English egalitarian institutions' and that beaches are 'places where differences are tolerated and eccentricities positively encouraged'. He suggests that beaches represent the most valued aspects of the English character, 'or the most agreeable facets of ourselves that we choose to elevate to national characteristics, i.e., our level-headedness, our sense of fair play, our stoicism, our anti-authoritarianism, our respect for individuality, our distrust of showy displays of intelligence or wealth'. He concludes that 'the seaside is easy to imbue with a mythic significance'. Published in 2010, Elborough's book was written before Carswell won Clacton and triggered the turmoil of Brexit and the national polarisation which ensued. Such an affectionate reading of English beach culture now sounds like a lost golden age; in a little over a decade, the tectonic plates of cultural self-understanding have shifted and a new national self-image has taken shape which is more aggressive, and ill at ease with itself.

Canvey Island is the seaside resort which defies all the geographic odds. An area of marshland, it sits beside the surging tides of the Thames, forty miles east of London. On three sides it is bordered by muddy tidal creeks: Holehaven, Benfleet and East Haven. The land is well below sea level, and its 45,000 residents live behind flood defences which rise above the height of their roofs, earth banked up against the concrete. The wall was raised four metres in the 1970s, and may have to be raised again as sea levels rise. Canvey Island lost fifty-eight lives in the great North Sea flood of 1953, and the risk of another such tragedy is a constant menace. The island's water table is high, so heavy rainfall and surface water can be as much of a threat as a breach of the sea walls. Current anxieties are focused on a roundabout where the routes converge and which could prove a bottleneck in an emergency evacuation. This implausible setting for a resort has not deterred the growth of Canvey, nor has the domineering presence of neighbours such as the gas distribution depot beside Thorney Bay, where families picnic on the sandy beach seemingly indifferent to the adjacent pale-green gas containers.

To the west of the island, caravan sites sprawl beside the rotting jetties, wharves and metal structures which stalk the muddy shores of Holehaven. Here, the concrete platforms are slowly cracking and crumbling as the weeds push through, the chain-link fences sag, and the tides deposit a thick ruff of plastic

amongst the scrub which edges the water. Teenagers gather on bikes to smoke and chat.

Every view is hedged by industrial infrastructure: the former Coryton oil refinery to the west, and oil storage tanks across the water on the Kent coast. Most spectacular of all, Canvey's modest beach is intermittently dwarfed by the vast freight tankers which glide past, perilously close to the indifferent windsurfers, on their way to the new Dubai-owned London Gateway port upstream. Canvey is intimate with this global traffic. I watched, incredulous, as one container ship majestically thundered into view, as tall as a tower block, blotting out the Kent coast. Beside it, silhouetted in a black wetsuit, a man on his hydrofoil surfboard cut through the water, dangerously close to its path, with the grace of a dancer.

Each time I visited Canvey Island, the promenade along the sea wall was busy: dog walkers, teenagers hanging out on the benches, kids skateboarding, families eating fish and chips, and people watching the world go past, literally. The shabby attractions – dodgems and amusement arcade – may have lost their appeal, but queues ran along the pavement for the fish and chip shops. On a sunny day, the beach was busy with sunbathers. A deep affection for Canvey was evident in the dedication plaques on each of the promenade's many benches; loved grandparents and parents were celebrated for their qualities – 'gentle', 'loving', 'warm-hearted' and 'deeply missed', and invariably mention was made of the pleasures they found on Canvey Island's seafront during 'many happy hours'. At one bench, posters offered a £500 reward for information about damage done to the benches. The place bristled with strong emotions, a passionate yet defensive love of the place.

By a paddling pool built into the sea wall, a mural portrayed the history of Canvey and the devastating 1953 flood. One vivid image showed a surging tide of water pouring over the sea wall on to the roofs of Canvey. No child growing up in the area can be unaware of its vulnerability. It's the stuff of a community's nightmares.

Canvey shares some of Jaywick's history as a resort for the crowded East End. Londoners originally came in search of cheap land for holiday homes, and then for permanent homes with space for a garden and fresh air, keen to escape the smog of the city. The island's history has long been one of refuge, after a community of 200 Dutch refugees arrived in the early seventeenth century, fleeing savage Spanish repression; many of the streets still bear Dutch names.

Canvey's hopes of becoming a significant seaside resort were dogged by its precarity. Floods in 1904 dashed the plans of one developer for a pier and six miles of glasshouses, exotic plants and fishponds. But one project which reached completion was the modernist architectural gem, the Labworth Beach Cafe, Ove Arup's tribute to the possibilities of concrete as a building material, streamlined to echo the passing ships. Its large windows were designed to double back to open the interior to the breeze, sunshine and water lapping a few feet away; this was a new seaside culture which revelled in the elements. More recent attempts to spruce up Canvey's seafront have offered a touch of unexpected domesticity to contrast with the neighbouring industrial infrastructure. The promenade along Thorney Bay opposite the gas storage depot has been painted pale blue (to offer some colour on a grey day) and decorated by large terracotta tubs of palms and geraniums.

It is 4.30 p.m. on a chilly November day and the light is fading, but there are still plenty of people on the seafront, along with a strong smell of chip fat; teenagers are necking on the benches in the dusk. I have been visiting the marshes of west Canvey and am in search of tea and chips. The tide is out, exposing soft sand. The lights flicker across the water in Kent and brilliant floodlights glow over the oil storage depots. Green and red navigation lights flash in the damp gloaming. I can hear the juddering power of the engines of a container ship before it becomes visible. First it sets the whole sea wall vibrating, then the pinpricks of lights appear of a huge ship.

Earlier in the day, I arrived at the seventeenth-century Lobster Smack Inn, which has survived high tides against all the odds, and now crouches well below the high dyke, a remarkable survivor amid the surrounding industrialisation. The inn is right by Holehaven Creek, and in the past smugglers could moor within a few feet of the inn. Allegedly it inspired Charles Dickens's Sluice House in *Great Expectations*. In the car park, I got into conversation with another visitor.

'I wouldn't live here – never.' When I ask why not, he roars with laughter, and then pauses a moment for emphasis. 'Well, put it this way, I used to work on the Essex highways for the council – and the councillors had brown envelopes stuffed under their beds. There's too many houses down here.' Every second sentence was a joke or a play on words. 'They get floods all the time,' he adds, laughing, and explains that he lives three miles away in a village where his wife's family have lived for centuries.

Climbing up on to the sea wall, I look over rows of static caravans between the storage depot and the sea wall. Some might find it an odd place for a holiday, but I can see the appeal. Down by the water, it is another world – of oystercatchers, curlews, and mud glistening in the pearly grey sunlight glowing through ragged cloud. The steady hum of traffic to the nearby shopping centre recedes, and I hear only the soft murmuring of the water slipping as the tide falls. As the waders pick over the wet mud, they are reflected in the glassy surface as if on a silvered skating rink. This was the Thames Estuary which entranced Joseph Conrad and inspired its role in *Heart of Darkness*:

> Forthwith a change came over the waters, and the serenity became less brilliant but more profound. The old river in its broad reach rested unruffled at the decline of day, after ages of good service done to the race that peopled its banks, spread out in the tranquil dignity of a waterway leading to the uttermost ends of the earth ... The tidal current runs to and fro in its unceasing service, crowded with memories of men and ships it has borne to the rest of home or to the battles of the sea.

I count seven supermarket trolleys encrusted with seaweed. A few other walkers are about, and when I ask one woman the way, she brims with enthusiasm. She lives nearby and this is her regular dog walk. 'I love it,' she says, waving her arms to encompass the mudflats, the industrial works across the water, and the cranes of London Gateway in the distance. 'I don't do pretty.'

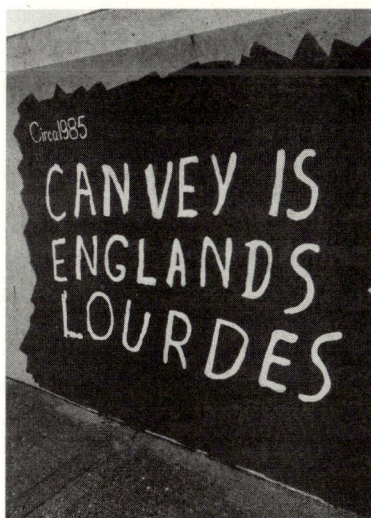

But in the misty sunshine, it is pretty. The brush and birch of the marshes are softly golden in the sun and the few boats are silhouetted. Even the storage tanks look striking in this milky autumnal light, faintly pink as the sun dips, and the dock cranes beyond are a crochet of dark grey against the sky.

A little further on, I meet Mike. His friendly face has an openness. He comments on the quantity of rubbish, rattling off a series of local community initiatives to pick up litter in other parts of the island, but this spot needs attention. He's a local councillor, he explains. After a few questions, he launches into a lengthy and riveting account of his election. The gist was that he had never thought to stand, but he was a 'Canvey man, born and bred'. He was a man who spoke his mind, and was not going to say what he was told to say by anyone. On election day he was at his father's bedside in hospital, and the news of his election victory came through. His friends were cheering on the end of the phone. He told his dad, who gave him the thumbs up, and, shortly after, died.

Mike is standing on the seaweed, the water a few feet away and the sun setting behind him, and I am gripped as much by how he told his tale as what he said. He was an actor playing all the parts of a play.

'I started my election campaign in very bitter circumstances.' He then embarks on an elaborate story involving a relationship, a fight and another man. 'He just laughed, he even spat in my face. He said, "You're just Canvey, you can't do anything." So the Canvey boy in me came out. He ended up in hospital. I'm not saying that's right, but it's what I did. So I'm down at Rayleigh police station – I didn't go to prison – and when I got out, a friend came to pick me up. He said, "Do you want a drink?" We were just passing a pub. And I said, "I don't mind if I do." We pulled in, and I said, "The car park is really busy," and he said, "Yes, it is, isn't it?" But I didn't think more about it until I went into the pub and everyone stood up, and cheered for me, and three ladies came over and asked if I would stand for the county council.'

He is beaming. Since then his political career has gone from strength to strength. Two issues he raised with the local MPs have now been mentioned in parliament, and one MP gave an hour-long speech based on his work.

'On one occasion, the posh Essex county councillors took me aside. I said to them, "I'm sorry, I speak my mind." They said, "Wait a moment, just hear us out." Then they told me that in just the short time I had been a councillor, I had made a difference. I speak my mind. Like when someone on the council said I was racist, I defended myself: I used the word "coloured" but I'm not racist, my parents were going to adopt a coloured girl. That's the word we grew up using, that doesn't make me racist. When the Assembly of God people arrived to live in Canvey – they are coloured – they asked me as a councillor to come and talk at their inauguration. We are all sitting there in the pews and when we get up, they all burst out, "Hallelujah!" Gave me such a shock! Anyway, I have to speak to them and I said, "I'll be straight, I speak my mind. You are very welcome on Canvey. Canvey has a wall all the way around, we're an island, and we have Canvey ways – they may not always be right, but that's Canvey for you. If you can get along with us the way we are, you'll have no trouble." And they stood up and gave me a standing ovation!'

The district of Castle Point, which includes Canvey Island, had the third highest vote in the country in favour of leaving the EU in 2016 – 72.7 per cent – narrowly beaten by two other east-coast districts, Boston and South Holland, which both border the Wash. The Brexit vote had a clear east-coast tilt; of the top ten highest Leave-voting areas, seven were on the east coast, and one other, Fenland, was within a few miles. The top twenty includes another batch of east-coast districts, such as Tendring, Hartlepool and Canvey's adjoining district of Basildon. London's eastern neighbours (including Tendring) accounted for six out of the top twenty highest Brexit votes. Douglas Carswell's claim that the Brexit revolution started in this corner of England has validity.

Rebellion has a long history here. Wat Tyler, the leader of the 1381 Peasants' Revolt is believed to have come from Fobbing, on Holehaven Creek, a few miles from Canvey; he led a march to London against a poll tax and serfdom, and lost his head in the process. The subject of independence has reverberated through history – Mike referred to Essex as the 'mainland' and argued that Canvey 'should go it alone', and it has its own Canvey Independence Party. This Essex coastal rebelliousness has its most outlandish manifestation on a Second World War iron naval fort in the sea at the mouth of the Thames. In 1967 Paddy Roy Bates, a former major in the Royal Fusiliers, moved into one of the forts, Rough Towers, which is seven nautical miles off Harwich, and just outside UK territorial waters. He declared independence and named his principality Sealand. Bates gave himself the title of 'prince' and issued Sealand's own stamps, coins, visa and passports. Over the last five decades, the Bates family claim they have defended it from the warships of several nations, including a joint German-Austrian raid in 1978. The Bates's exploits on their crumbling Second World War fort makes the yearning for independence of the landlubbers on the nearby coast seem moderate by comparison.

This coastline may be densely populated, but it still has a wild character tucked into the mud creeks. It has continued to defy the political power of London. South Essex is a place with a lot to say about itself. Fiercely held traditions of pride and loyalty feed into an intense patriotism. It is proud of its prosperity; South

Benfleet, perched overlooking Canvey Island and the Thames
Estuary, is a neighbourhood designed to impress. Each mansion is
more grandiose than its neighbour. This is not the kind of wealth
discreetly hidden behind trees and hedges; it announces itself to
passing traffic: Grecian columns, chrome, plate glass, large garages
and sweeping drives on which to park the expensive car.

There are many aspects of Essex which are upfront and
direct, so it can be easy to overlook what has been obscured,
even hidden. Signs will guide you to the birdlife, and the heritage sites, but nothing explains the industrial infrastructure
which dominates the area. Oil and gas are a 'submarine industry', comments the energy expert Professor Paul Stevens, its
workings often concealed behind fences. Infrastructure springs
up, gets sold and is then closed down by distant corporations
with bland names, all within a generation or two, in a disorientating sequence of boom and bust. What's left behind is the
collapsing, derelict infrastructure, evident in several places along
this Thames coast. Further upstream from Canvey, between
Tilbury and Stanford-Le-Hope (where Conrad once lived) one
walks past the spaghetti of rusting pipes surrounded by structures refusing to explain themselves.

Essex's coastline is one of beguiling juxtapositions, thick with
histories, many of which have been neglected, washed away, forgotten, or were too illegal to ever be recorded. It's a place where
people have always been able to hide in its creeks and mudflats.
Frinton-on-Sea's genteel avenues and pretty beach huts, the buzz
of Southend's casino and palm-tree strip, and Canvey's sea wall
are just the most recent overlay. It's a coastline which hooked me
in, and left me wanting more: to follow more of the crazed glazing of creeks, mud and crumbling industrial infrastructure, and
discover more of the glamour, eccentricity and fun. I had to pull
myself away, Kent was calling. I was on my way south.

4

Margate to Folkestone

Kent

The eastern part of Kent known as the Isle of Thanet has not been an island since the Middle Ages, when the River Wantsum silted up and the surrounding low-lying land was drained. But this stretch of coastline, which includes a cluster of famous sea-side resorts – Margate, Broadstairs and Ramsgate – is still known by this outdated geographical feature. The Ordnance Survey map also indicates a Plumpudding Island, along the coast from Reculver; long after these islands have ceased to exist, they continue in phantom form as geographical possibilities, and local historians insist that this island history is critical to grasping the area's very particular character. Thanet and the adjacent south Kent coast has always been a crossroads, facing east to Belgium and Holland, and south to France. The finger of land juts out into the North Sea, into one of the busiest shipping lanes in the world.

The narrowness of the Channel gave prospective invaders an advantage; most famously, the Romans landed here, not far from Ramsgate, but also the Angles, Saxons and Normans. Ever since, every war with Europe has provoked intense anxiety. For centuries, elaborate fortifications have been constructed, such as the Tudor fort at Deal (1539) and, further down the coast, Dover Castle (the first iteration was built in the eleventh century). In the Napoleonic era the threat of invasion prompted the erection of 103 Martello towers around the English coast between 1805 and 1812, seventy-four along the Kent and Sussex coasts, built to withstand cannon fire, as well as the huge undertaking of a twenty-eight-mile canal from Hythe, near Folkestone, to Cliff End, near Hastings, to act as a barrier to an invading force on this stretch of coast. Shakespeare's lines in *Richard II* were part wishful thinking: 'This fortress built by Nature for herself/ Against infection and the hand of war'. In fact, Nature was not sufficient, and where steep cliffs gave way to low-lying beach, immense resources had to be invested for centuries to protect the nation from the neighbours just twenty miles across the Channel. Before this coastline began to welcome visitors, its primary focus was defence.

Margate's proximity to London gave momentum to its development as a beach resort and, along with Scarborough and Brighton, it was well established by the middle of the eighteenth century as one of the early pioneers. Londoners were able to take passenger ferries down the Thames, and during the early nineteenth century their numbers soared. The modest villages and fishing ports of Thanet swelled into vibrant resorts, each with its own character and reputation, and yet only a few miles apart. The infrastructure for pleasure and war sat side by side, and, more so than on any other English coastline, multiple national mythologies have accumulated along the sixty-mile stretch around Thanet and down to Dungeness: contradictory, argumentative and happy to speak for the whole nation.

*

When the first Covid-19 lockdown eased slightly in the May of 2020, and the prime minister Boris Johnson informed us that we could go to the beach to take our hour of exercise, sea resorts braced themselves for the invasion. But on my visit to Margate on a sunny weekday, a nervous nation had chosen to stay home. Apart from a group of teenagers playing wheelies, smoking, and laughing raucously, the town is eerily quiet, and the quaint cafes, pubs and art studios are all closed.

In one cafe window, the empty tables sit expectantly waiting for customers, their Union Jack tablecloths splashed with sunlight. Next door, a banner advertises Bombardier beer, displaying the image of a whiskered eighteenth-century soldier and the St

George Cross flag under its slogan, 'Glorious English'. Here is an edgy nationalism, a British tradition of military endeavour, bonhomie and ale, but this national crisis will not be resolved by the armed forces – despite the frequent use of military metaphors – nor by bonhomie, given the requirement for social distancing, so what's left is beer, only available from the supermarket. Everything is out of joint. The large buildings of the Turner Contemporary art gallery, prominent on the seafront, are silent, the car park empty. A few fishermen have cast their lines out at the end of the silent pier. I sit on Margate's golden sand, the sea sparkling in the sunshine, virtually alone; there must have been few comparable occasions in the last two hundred years.

Margate holds a prominent place in the national geography, a pioneer of both the rise and the decline of the English seaside resort. Signs in the historic centre chart the town's inventions: the first to offer a commercial seawater bath in 1736, the first reference to a boarding house in 1770, to a seaside hotel in 1774, seaside theatre in 1786, first seaside donkey ride in 1790, first hospital to pioneer the use of open-air and salt-water treatments, first seaside use of deckchairs in 1898. Its reputation as *the* Cockney seaside resort and its rail and road links to London still ensure that thousands of visitors fill the beach and pack the arcades and the regenerated art deco Dreamland amusement park all summer. Behind the busy seafront, however, the town hit hard times in the 1980s; the visitors were still coming, but, as in other resorts, they were not staying in the dozens of small hotels. As the town's fortunes plunged, it became as famous for the extent of its seedy decline as it had once been for its popularity. Along the way, both in its heyday and at the worst point of its decline, Margate has served as a screen on to which artists, poets, writers and film-makers have projected their preoccupations.

Like thousands of other Londoners in the early nineteenth century, the artist J. M. W. Turner, one of Margate's most famous visitors, travelled there by steamboat down the Thames. He was in company; by 1815 there was a regular service from London Bridge, which was soon carrying 40,000 passengers a year, and by 1840 an estimated 100,000 took the eight-hour journey. Travellers played backgammon, chess and draughts on the boats, as described by the young Charles Dickens in his first published work, *Sketches by Boz*. Britain was forging a new relationship with the sea, and the thousands of steamboat passengers marked a new maritime intimacy; sailing had become a form of entertainment rather than stoic endurance. Turner loved Margate and visited regularly. He used its shoreline in his work to capture a nation's focus shifting out to its edges – its pebbly beaches, cliffs and surrounding seas. 'While the inlands of England have been so hackneyed by travellers and quartos, the Coast has hitherto been unaccountably neglected, and, if we except a few fashionable watering places, is entirely unknown to the public,' wrote the critic Richard Ayrton in 1814, expressing his surprise that this was so, since 'so many associations flattering to our [national] pride are connected with every view of our seas and shores'. Turner and his contemporaries, John Constable and printmaker William Daniell, inspired – and fulfilled – a new appetite to know England's coastline, travelling extensively and producing the paintings and engravings which introduced many to their country's edges. Their work 'took on the form of a hymn to the insularity of a nation that had just overcome terrible trials [the Napoleonic wars] . . . the huge labour of artists between 1800 and 1840 was proof of the prestige of the seaside', writes Alain Corbin in *The Lure of the Sea*. The result was that the coast was 'as much a place for imaginative projection as actual visitation', and became part of the 'daily cultural life of the nation . . . a limit and a horizon, the coast offered a way of looking at once outwards and inward'.

Famously, Turner told John Ruskin that the 'skies over Thanet are the loveliest in Europe', and painted more than a hundred works depicting this Kent coastline. Amongst this flat marshland, Turner shifted his eyes to the sky and out to sea, to paint mist, light, clouds and the movement of water. Daniell's prints of Thanet effectively record a process of property development as the Kent towns expanded, but Turner offered a grander vision: his sea was awe-inspiring and majestic, no longer just a place of dread but a source of inspiration; as his contemporary, Lord Byron, wrote in his poem, *Childe Harold's Pilgrimage*, the sea was 'boundless, endless and sublime,/ The image of Eternity – the throne/ Of the Invisible'. Turner painted air – one French critic described it as 'active nothingness' – clouds and birds. His paintings left some critics baffled; Hazlitt commented that Turner's work were 'pictures of nothing and very [a]like'. But this dialogue with nothingness 'speaks tirelessly to the Romantic soul', writes Corbin. 'The Romantics made the seashore the favourite spot for self-knowledge.' They gazed at the horizon, which had taken on new meaning. Previously, 'the awe-inspiring dimension had been vertical', points out John Gillis in *The Human Shore*, but for the Romantics the horizon became the 'frontier of the imagination'. As Britain industrialised, 'the sea alone remained unchanged', and became 'the repository of all that had disappeared on land', the last place of mystery and wilderness.

For the British 'the ocean [was] at once their nation's proper domain, the central geographical theatre of their age, and the main natural force shaping their world', suggests historian Samuel Baker. This was the theme Turner explored, providing a visual vocabulary for this formative period of national identity, by choosing as subject matter the different maritime worlds of fishing, naval warfare, defence and the new tourism. He painted the struggle of fishermen hauling a boat in amidst treacherous waves, and the poignancy of an ageing warship coming home to dock. 'Of all his contemporaries [he] was the landscape artist who embraced the modern world and industrialisation and [grasped] that a key characteristic of that modernity was Britain's relationship with its sea – as a defensive cordon, as a stage for heroic endeavour, expansion and pleasure,' writes art historian Christine Riding.

A century later, in 1940, Britain faced the threat of invasion, and the poet John Betjeman returned to Margate to illustrate the country's existential crisis. Hitler had invaded France, German divisions could be seen on the French coast from Kent, and Britain's future lay with the success of the young air pilots in the skies overhead. Barbed wire and concrete defences were being hastily erected along beaches, mines were being laid, areas were fenced off to civilians, and seaside towns evacuated. Winston Churchill famously warned that the beach, that icon of Englishness, could become its battlefield. Betjeman's poem was already nostalgic for the lost ordinariness of the pre-war resort.

> *How lightly municipal, meltingly tarr'd,*
> *Were the walks through the lawns by the*
> > *Queen's Promenade*
> *As soft over Cliftonville languished the light*
> *Down Harold Road, Norfolk Road, into the night.*
>
> *Oh! then what a pleasure to see the ground floor*
> *With tables for two laid as tables for four,*
> *And bottles of sauce and Kia-Ora and squash*
> *Awaiting their owners who'd gone up to wash –*

Mournfully Betjeman concludes:

> *Beside The Queen's Highcliffe now rank grows the vetch,*
> *Now dark is the terrace, a storm-battered stretch;*
> *And I think, as the fairy-lit sights I recall,*
> *It is those we are fighting for, foremost of all.*

Margate is emblematic of all the love, history and decay of the English seaside resort. Fittingly, Graham Swift chose Margate as the site of pilgrimage in his novel *Last Orders*, in which four old friends from Bermondsey in south-east London fulfil their friend's last wishes by scattering his ashes off the end of the pier. For these East Enders, the famous medieval tradition of pilgrimage across Kent is not now to Canterbury but to Margate. Swift explores the complex relationship between east London and its seaside hinterlands – a place of dreams, brief escapes and moments of enchantment snatched from hard lives. The journey of his characters veers between pathos, comedy and poignancy; the ashes are kept in what looks like a jar of instant coffee, one character observes. Swift's Margate is a place of past pleasure but also boredom. It combines the mundane and the tragic.

In the late 1990s and early 2000s, the portrayal of Margate became bleak. The film *The Last Resort* (2000) portrayed Margate's devastating decay through the plight of a Russian asylum seeker and her young son, while the artist Tracey Emin wrote in her autobiography of its poverty and social breakdown in the 1970s and 1980s, describing how that played out in lives wrecked by drugs, drink and sex abuse. Galvanised into action, the council launched a bold, culture-led regeneration project to turn the town around; with the opening of the Turner Contemporary art gallery in 2011 and a revitalised Dreamland in 2015, Margate has been celebrated as a successful model of how to reverse seaside resort decline.

Despite the success of Margate's regeneration, parts of the centre of the town are still amongst the most deprived in the country; it has three areas in the top half per cent on the Index of Multiple Deprivation. In common with other towns on this Kent coast, such as Ramsgate, Dover and Folkestone, small areas of deep poverty have persisted, unaffected by programmes of investment and renewal. The average weekly wage of £528 in the district of Thanet (which covers Margate and Ramsgate) is considerably lower than the south-east average (£614) or the national figure (£571). More than a third of all children in Thanet were defined as living in poverty in 2019, compared to 21 per cent in the more prosperous parts of Kent such as Tunbridge Wells; the fact that 70 per cent have at least one parent in work reflects one of the main drivers of coastal poverty: the dominance of low-skill, low-paid seasonal work. Even before Covid, unemployment was nearly twice the national rate in these coastal towns. This pre-existing economic fragility was exacerbated by the pandemic and Thanet saw an increase of a third in Universal Credit claimants during Covid. Other indicators of hardship include the high rate of insolvencies in 2020 in Thanet and Dover, which were twice those inland, while repossessions on the coast were ten times the number in prosperous Tunbridge Wells in the last quarter of 2021. Youth unemployment in Thanet was 10 per cent in December 2021, nearly twice the national average, and three times that of nearby Canterbury.

Cliftonville, in east Margate, was ranked as the fifth most deprived area in the whole country on measures of education, skills and training for children and young people; 16 per cent fail to achieve five GCSEs A*–G in Thanet. Providing good-quality education in a context of deprivation is always a challenge, but in recent decades it has become particularly acute in coastal and rural communities with little diversity. They struggle to recruit good teachers, and too often a culture of low aspiration sets in, which further exacerbates recruitment; the national teacher shortage is predicted to worsen in the coming years and coastal and rural schools will suffer disproportionately. The issue is well known, yet successive governments have failed to tackle the most egregious aspect: the longstanding unfairness of the funding formulas which ensure that inner-city schools have significantly more money. In 2022 Kent received £4,367 for each primary-school pupil and £5,679 per secondary pupil, compared to the London Borough of Hackney, which received £6,356 and £8,501 respectively. When multiplied by several hundred pupils, this disparity amounts to considerable sums for each school, affecting teacher–pupil ratios.

The education challenge in Thanet is compounded by a further factor: a disproportionate number of looked-after children are sent to the area. This trend began in the 1990s, when former hotels were converted into children's homes, and fostering provided an important income for many households. For London boroughs whose demand for placements exceeded supply after they had sold off their own children's homes due to budget pressures, Thanet and other coastal Kent towns, like Folkestone, seemed a good option. Costs were reasonable, the holiday associations made them popular, and some children needed to get away from abusive or dangerous environments. But the numbers kept rising and, inevitably, they added to the demand for health and education services. For several years now, Thanet District Council has been lobbying government to limit the numbers of looked-after children sent to the area. In 2018 a headteacher complained vociferously and threatened to ban them from his school, provoking fierce local debate over whether this was a form of Nimbyism – Not in My Backyard – or a legitimate complaint. The issue reached parliament, and the government minister conceded that 45 per cent of Kent's looked-after children were from outside the county.

In addition, as the numbers of migrants crossing the Channel in small boats increased, Kent found itself providing a home for Unaccompanied Asylum Seeking Children (UASC), and many ended up in Thanet. In 2021 Kent council said its services for UASC were at breaking point for the second time in less than a year, and refused to accept any more, saying the number had reached 'unsafe levels'. The county had over 400 UASCs, when the government had agreed the number should not exceed 231. Furthermore, many remain in Kent after reaching the age of eighteen and, understandably, still need support; the council estimated the total in 2021 at 1,100 under-twenty-fives. It issued legal proceedings against the home secretary for failing to enforce the distribution of children arriving in Dover more widely across the country. The council argued that the children are disproportionately male, often older than they claim to be, and have little English, putting added strain on the education system.

Kent is a county of extremes, with some London commuting areas among the most prosperous in the south-east, and a deprived coastal hinterland of Swale, Sheppey and Thanet. Within that pattern, a micro pattern of stark inequality, evident even within the few square miles of Thanet itself, is reflected in shocking variations in life expectancy. East Margate is only a few miles from the outskirts of prosperous Broadstairs, yet the gap in life expectancy between the two towns is 8.5 years lower for men in Margate and 10.2 years lower for women. Mortality among under-seventy-fives is nearly 40 per cent higher in

Thanet than in the rest of the south-east, and its suicide rate is
50 per cent higher. Thanet has one of the highest rates of drug-
related deaths in the country, three times the national rate. A
Kent-wide report by the county council admitted that 'health
inequality was getting wider'. Locals know it; a woman I talked
to on the seafront in Ramsgate grimly concluded that over the
twenty years she had lived in the town, 'The rich have got richer
and the poor have got poorer.'

Like Jaywick, the Thanet seaside towns have served as
London's place of last resort for people in need of a new start.
Fond memories of a childhood holiday combined with relatively
cheap housing has been a draw for many who have had troubled
lives. The services struggle to cope, observes the sociologist
Sarah Cant from Canterbury Christ Church University, who
undertook a research project for Thanet District Council on the
work of a multi-agency taskforce of police, social workers and
the then Department of Work and Pensions.

'The taskforce set up a process of visiting a street at a time,
with representatives from all the agencies proactively knocking
on doors. It was shocking to find what was behind those doors:
cuckooing [taking over the home of a vulnerable person as a
base for illegal drug dealing] and prostitution were going on in
close proximity to the houses bought by "Down From London"
migrants, drawn to the cheap houses, the edginess of the area,
and the artistic legacy. In these same streets there were homes
for ex-convicts, hostels for women escaping domestic violence,
and children's homes; the problems feed off each other, and
young people are easily sucked into county lines of drug dealing.

'It was Farrow & Ball paint and wooden shutters on one house,
and next door a dilapidated House in Multiple Occupation, with
families struggling with overcrowding and unsafe living conditions. There are so many looked-after children sent to Thanet,

and the council has no control over the numbers arriving. Some local schools report having to manage complex needs with no additional financial support.

'The taskforce had such energy as they tried to make things better, but these coastal towns have limited capacity to cope. They already had challenges of their own, with poor transport links, and little economic diversification away from tourism. The social workers and police were running up the down escalator, constantly firefighting successive waves of problems – new drugs and new vulnerable residents and new migrants to house and settle.

'There was a lot of horrible racism, and people talked of how Margate had gone down the pan; many making observations along the lines of "I'm not racist but if it wasn't for the Eastern Europeans or the Roma or the drug addicts ... " But those who said that didn't live in the poorest areas or even visit them.

'You could paint a picture of Margate as part of "left-behind Britain", but, notwithstanding the racism, I also see it as a place of great resilience, where community and hope flourish and where there is conviviality in the most unexpected places. One recovering drug addict I interviewed had taken it upon herself to plant crocuses near her hostel, and was grateful for all the support the taskforce had offered her. I heard how Roma women picked up the needles every morning so that the kids could play safely in the playground. There were attempts to build bridges between communities and to tackle the division and racism.'

At the turn of the millennium, there was widespread recognition that the task of regeneration required big and bold investment; Thanet District Council, Kent County Council and the Arts Council England all contributed to a multi-million-pound budget to build the Turner Contemporary art gallery. The former director, Victoria Pomery, remembers arriving in 2001 to be interviewed for the job and being both shocked by the poverty and shabbiness and thrilled by the town's beauty and potential.

'The idea of a major modern art gallery was very brave – the Liverpool Tate had pioneered how culture could lead regeneration, and Tate St Ives had shown how successful it could be – but Margate was on its knees. It had no permanent collection so the offer would need to be different. It was a small town and the vision was big – a world-class gallery. The decision was that it had to match Margate's past; it couldn't be mediocre. It helped that the architect was David Chipperfield, and he produced a brilliant design.'

Pomery and her team had a huge task to persuade local people that the gallery would be a success. 'They wanted a new wing on the hospital and a new school. It was a long journey to change people's mindsets. We had to get people onside, and we talked to everyone.'

Gradually the project gathered momentum and investors started arriving in the town ahead of the Turner Contemporary opening in 2011. Artists in search of cheap studio space moved down from London. Boutique hotels opened, and the number of visitors to the gallery increased every year, so that by 2019 it had reached just under half a million a year, far exceeding original forecasts.

'The change was tangible. It's been a huge success,' reflected Pomery – she stayed nearly two decades before moving in 2021 to a new job – but she admits she is troubled by some aspects of the project's legacy. 'There are still high levels of deprivation. Local people said to me, "The Turner Contemporary is such a success, but my children cannot afford to buy a house here now." There's a thin line between regeneration and gentrification, and

I still haven't resolved a way through that. The arts are used for a placemaking agenda which can end up forcing out the very people, such as locals and artists, that you want to keep. I don't feel proud of the house prices going up, but I am proud of the civic pride and educational attainment in the town – culture can help. The gallery was free and had a big education and learning programme.'

In the first five years alone, two million visitors came to the Turner Contemporary, and nearly half came to Margate specifically to visit the gallery, spending an estimated additional £8 million annually in the town. In 2015 the iconic art deco Dreamland amusement park was reopened after a major regeneration project (the relaunch was dogged by problems, but after they were resolved in 2017, it attracted record visitor figures until the pandemic in 2020). Pete Doherty of the band The Libertines opened a quirky hotel, and Margate-born Tracey Emin moved back to the town to build a studio. Margate had become cool, adding to the lure of its golden sands, breath-taking sunsets, the sea and the quaint eighteenth- and nineteenth-century terraced houses in the centre. By 2015 the town had the fastest-rising house prices in the country outside London, with a year-on-year rise peaking at 24 per cent. Locals were particularly aggravated by the rise of Airbnb: properties were bought up as second homes for holiday rental income and often left empty over the winter. A campaign group called for restrictions, such as those in London, where a property can only be let for a maximum of ninety days a year. They argued that Airbnb had crippled the long-term rental market and had raised prices, leaving a shortage of affordable housing. They have a point; in March 2022 Airbnb offered 300 properties to rent, but my search for long-term rentals produced only twenty-three flats, starting at £600 per month for a one-bed.

The pandemic pushed property prices up even further as Londoners looked to move out of the city for more space; one property survey recommended the Thanet towns as the most desirable locations for those wanting to leave the city, given their relatively easy access to the capital and – in comparison to London – low prices. The influx of money has created tensions and cliques known by their acronyms: DFL, DFA (Down from Anywhere) and DFSE (Down from the South-East).

'The creatives like the grimy edge but get upset when a drug dealer vomits on their front door,' pointed out Cant. 'A lot came and left quite quickly, but there is now a significant body who have stayed. Has the arts-led regeneration worked? It depends on your lens: has it led to Airbnb and boutique shops? Yes. Has it led to an increase in house prices and more visitors from London? Yes and yes. But if you ask who benefits, it's a classic story of gentrification. The well-meaning artists want to get kids into art and they run projects, and the kids hang out in the Turner gallery, because it's warm and the WiFi is free, until they are booted out, but no local people go to the gallery to look at art. The creatives and the locals hang out in their own bubbles. The white working-class mums were in awe of the "creatives" mums at the school gate: "They are so beautiful, so stylish," they told me, but they didn't talk to them.

'Margate council now has powers to restrict the number of HMOs [Houses in Multiple Occupation], and they can insist landlords do repairs and have issued orders to that effect. But there have been some dreadful landlords who bought up swathes of property and let them fall into a despicable state – and then they are sold to creatives who do them up. I was shown round one such renovated house and the owner proudly told me that each room had been named after an artist; she added that before they bought it, each room used to house a family. "I do wonder where they have gone," she said to me.'

Pomery was particularly proud of an initiative which managed to bridge these deep divides. The curatorship of a major exhibition was handed over to members of the community, and for three years around thirty Margate residents were involved in a process of discussion and collaboration for an exhibition on T. S. Eliot's *The Waste Land* in 2018. Artists as diverse as Edward Hopper, Tacita Dean and Philip Guston participated in the widely praised show. 'For me it was a turning point – it brought together senior academics and people who had never read *The Waste Land*. But the hardest question is how to ensure the transformation benefits everyone,' admits Pomery.

The artists' studios have brought with them yoga classes and smart coffee shops; in January 2022 Tracey Emin announced that her art studio in Margate will become a museum of her work after her death and that she is building thirty studios for young artists. A poll in Thanet on whether the DFLs and DFAs benefited the area reflected the deeply held divisions, with 226 voting in favour and 164 against.

'The houses look nicer, tidier and have been beautifully restored, but, despite the pace of change, it doesn't seem to be touching the deprivation,' said Cant. In the book she is writing she tells 'The Tale of Two Cafes': 'a greasy-spoon cafe and one frequented by creatives, and they exist in parallel universes. The creatives will visit the greasy spoon, but not the other way around. It feels uncomfortably like colonisation. I found the categorisation of American sociologist Arlie Russell Hochschild helpful: people who have got nowhere; people who have got somewhere; people who can go anywhere. Creatives fall into the last category, but the Margate working class feel they don't have somewhere, it's been eroded beneath them.'

Six months later, in another lockdown, I was back in Margate to cycle the eight miles along the seafront to Ramsgate. I started

out at the famous shelter at Nayland Rock, where the poet T. S.
Eliot sat in 1921, on leave from his job at Lloyds Bank. He had
come to Margate to convalesce after a nervous breakdown. His
doctor had told him not to write, but he couldn't help himself,
and drafted the third part of his modernist masterpiece on the
catastrophe of the First World War, *The Waste Land*. He sat on
the bench, watching the beach where children were playing and
wounded First World War veterans were painfully doing their
rehabilitation exercises.

> *On Margate Sands.*
> *I can connect*
> *Nothing with nothing.*

The sky was overcast and still, the sea a strange grey-green.
It was a cold day, but there was no hope of a warm cafe, as they
were all shut. I left the town behind me, skidding over the sand-
strewn concrete. Stubby chalk cliffs loomed above, streaked
with grey and green mould, and frequently patched with brick,
stone and concrete, marking different eras of construction and
repair. Below, the sea wall had been blackened by the sea and
coated with seaweed, emerald green shaded into a dark bottle
green, the sharpest colour in this palette of greys and browns –
apart from the intermittent splash of painted graffiti. This outer
edge of Margate was part urban edge land – the intimidating
bulk of Foreness pumping station defining it as a place of industrial infrastructure – and part seaside promenade, dotted with
bunches of dying flowers tied to the railings as memorials. Nine
metres above, green lawns edged the cliffs, and teenagers hung
out in shelters to smoke dope. The only connection between
these two seafronts were the steeply sloping pathways cut into
the chalk. In lockdown England, it was busy; exercise was the

only permitted reason to leave home, so a nation had wrapped up in heavy coats and scarves to perambulate the coast like medieval pilgrims. The sand of Margate gave way to a messy patchwork of rumpled chalk, pocked with small pools and dark-brown seaweed. It looked like an afterthought, the leftovers of a gigantic construction project: England.

Between Margate and Broadstairs sits the suburb of North Foreland, discreetly perched on the tip of the Isle of Thanet, only a few degrees short of the most easterly point of the UK (at Lowestoft Ness). It has been the site of a succession of light-houses since 1636, and overlooks the steel-grey seas and the procession of commercial vessels which pass through the Dover Strait, on average four hundred of them every twenty-four hours. As a bleakly exposed vantage point, it bears a resemblance to a bridge over a motorway. But in its favour is an unobstructed view of the sunrise. North Foreland was in the top decile of the *least* deprived areas of the UK in 2019. The contrast between the shabbiness of Cliftonville, three miles away, and this gated private estate was striking. The cars were expensive, the secu-rity tight, and the clifftop houses had huge plate-glass windows. Plots with planning permission started from £450,000 (up the road in Margate you could buy a block of three flats and a shop in the historic town centre for £650,000). A new development within the estate, Ocean Drive, offered a fully gated compound (as if one set of gates was insufficient), where £1.2 million bought a three-bedroom apartment with a sea view and private steps down to the narrow pebbled beach. And these were not just any steps: they were those which John Buchan incorporated into his novel *The Thirty-Nine Steps*. In August 1914, on a holiday to the area, Buchan visited a cousin renting a villa at North Foreland with a set of steps zigzagging through two shafts and three tun-nels down to a private beach.

The wealth of nineteenth-century Broadstairs may be less ostentatious than that of North Foreland, but its quaint charms are comparably expensive. It is famous for its association with Charles Dickens, who had a house there and visited regularly; the novelist said he used the sea view as inspiration for his writing. Amongst his earliest writings is a satire of the newly fashionable Thanet coastline, 'The Tuggs's at Ramsgate', published in 1836. The Tuggs are a family of Cockney greengrocers who come into an inheritance and decide they 'must leave town immediately', and that means the seaside, but they are undecided about where to go: Gravesend is 'too low', and Margate 'worse – nobody there but tradespeople', so they settle on Ramsgate:

> The sun was shining brightly; the sea, dancing to its own music, rolled merrily in; crowds of people promenaded to and fro; young ladies tittered; old ladies talked; nursemaids displayed their charms to the greatest possible advantage; and their little charges ran up and down, and to and fro, and in and out, under the feet, and between the legs, of the assembled concourse, in the most playful and exhilarating manner. There were old gentlemen, trying to make out objects through long telescopes; and young ones, making objects of themselves in open shirt-collars; ladies, carrying about portable chairs, and portable chairs carrying about invalids; parties, waiting on the pier for parties who had come by the steam-boat; and nothing was to be heard but talking, laughing, welcoming, and merriment.

In this unfamiliar environment, the Tuggs are out of their depth, anxious about their social status and keen to improve it. On their arrival in Ramsgate, they pay over the odds for inferior accommodation with 'a bay window [from] which you could obtain a beautiful glimpse of the sea – if you thrust your body out of it'. Dickens pokes fun at the new Romantic appetite for the sea: 'the calm sea dashed against the tall daunt cliffs with just enough noise to lull the old fish to sleep without disturbing the young'. The Tuggs end up defrauded of their fortune by a couple they had ingratiatingly befriended, believing them to be their social superiors. The seaside resort in nineteenth-century Britain was a new form of public space where different classes could bump against each other, and status might not be evident or could even be deliberately and dangerously concealed, the reader was reminded. Dickens's contemporary, the satirical cartoonist George Cruikshank, depicted an aristocrat on the promenade in Brighton horrified to discover that a well-dressed fellow visitor was his own tailor. Historian Christiana Payne points out that this was one of the few places 'where classes mingled and anonymity enabled unexpected encounters'.

The development in the early decades of the nineteenth century of these new public spaces was radical; first came the invention of the promenade and the pier, next came the beach, which ceased to be simply a place for exercise (walking, riding and bathing) and evolved as a place for entertainment, sitting about on chairs, and play. The coast became a 'zone of cultural interchange', argues the literary critic Valentine Cunningham, as it attracted an eclectic combination of poets, geologists, painters, medical men, salesmen, entertainers, holidaymakers and invalids. The beach's appeal has become such an accepted convention that it is hard to imagine how revolutionary it originally was. The mix of men and women from different social classes gave rise to embarrassment as well as sexual frisson; despite segregated beaches and the use of bathing machines to conceal women bathers, it was possible to glimpse that rare thing in Victorian society – bare flesh. Many men bathed naked, while women wore long shifts which clung to their bodies; contemporary observers were horrified to note that both men and women unashamedly used binoculars and telescopes. The *Observer* reported in 1856 that in Margate and Ramsgate, as the women lay in the surf to bathe, their shifts rode up, exposing their bodies, and the gentlemen came with opera glasses to look, while many women went to the men's beach for the same purpose, and 'they calmly looked on without a blush or a giggle'. Segregation and stricter emphasis on bathing costumes ensued in the moral panic, but the beach had been established as a space for spectacle, and a tradition was established lasting to this day in which part of its pleasure is people watching without much inhibition.

One of the most famous paintings of the nineteenth century, *Ramsgate Sands* (1852–4) by William Powell Frith, captured this invention of the beach. A huge commercial success as a print, the painting was bought by Queen Victoria to hang in her new seaside retreat, Osborne House, on the Isle of Wight. The crowded beach scene is crammed with detail: children play in the surf, elderly gentlemen read newspapers, their wives are knitting, seated on stiff, upright chairs. It is as if the private domestic spaces of garden and sitting room have been turned out on to the sands. The picture captures the hawkers and entertainers who have spotted a new market, crowding among these sedentary customers to sell their wares and play their instruments. The mixing of social classes brought new dangers; the drama at the centre of the painting is a kneeling man who has caught the attention of a pretty young woman, while her mother looks on suspiciously. Frith captured what is now an established paradox of the beach: it is a public space where one can also be private; self-absorption can switch to casually observing the intimate comings and goings of beach neighbours – squabbles over the picnic, lost costumes and suntan lotion – before withdrawing again. Frith's groups are self-contained, each preoccupied with their own domestic routines.

The sand is only just visible on the crowded beach, and some of the upright chairs are standing in the surf: the liminal interface between land and water had been tamed and was now ignored. Nature and its threats have become almost irrelevant in this version of the seaside – except for the sun. To protect themselves, men wear broad-brimmed hats and hold umbrellas over their womenfolk, shawls and bonnets are pulled well down over ladies' faces to shield them. Even the toddlers are in bonnets. Bodies are elaborately covered with an extraordinary quantity of clothing: ribbons, shawls, ruffles, crinolines, embroidered bodies, flounces, cuffs and petticoats, neckties, top hats, bonnets, stiff collars and handkerchiefs. (A small army of mill workers, haberdashers, needlewomen, tailors, laundry women, ladies' maids and valets would have been required to assemble this crowd for their day on the beach.) In the background, Frith painted the imposing new buildings of the fashionable resort – a regency terrace, a neo-classical building, an obelisk – a mix of architectural styles which served as an appropriate stage for the visitors. The artist recognised that seaside architecture had already established itself as distinct, designed to indulge the visitors' tastes for fantasy, delight and the exotic. Some of the surplus building materials have not yet been tidied away; the seaside was still a novelty.

Frith's approach was artist as documentary-maker, and his work was immediately appreciated as such; the *Art Journal* of 1854 praised the painting 'as a memento of the habits and manners of the English "at the seaside" in the middle of the nineteenth century'. A Victorian audience was satisfied at their happy portrayal in the modern world that they were creating. Frith's interpretation was a far cry from Turner's grand national narratives of sea and empire; here in Frith's Ramsgate, the seaside, and more specifically the beach, has been domesticated for family life. The beach was no longer primarily a working place, nor a form of wasteland, or even a potential battlefield; it had become *homely*, tidy and safe.

Four years later, another painter, William Dyce, was also inspired by this stretch of Kent coast and produced a very different interpretation of England's deepening love affair with the seaside. *Pegwell Bay, Kent – A Recollection of October 5th 1858*, depicts the bay to the south of Ramsgate. In contrast to the earlier painting, it has only a few figures, dwarfed by the stretch of shore and cliff on a chilly evening as the sun is setting. In the foreground three women are exploring the rock pools. A small boy looks out to sea, his spade at his side. In the middle distance a group of workmen appear to be gathering stone, and in the far distance a few fishermen work on their nets. All the human figures are isolated, none look directly at us. The cliffs are painted in exacting detail, showing the strata of the chalk; in the sky a trace of Donati's Comet is visible (confirmed by the date in the painting's title). Human beings are fragile entities in a vast cosmos, their lives a mere blink in the passage of deep time. A year after the painting was produced, Charles Darwin published *On the Origin of Species*, unleashing the huge theological and scientific debate which profoundly challenged the Victorian complacency and self-satisfaction evident in *Ramsgate Sands*. The two paintings represent opposing and overlapping experiences of the Victorian seaside, a place of modern pleasures, but also one which presented the greatest questions of faith and science.

At the time Dyce was painting, the fashion had taken grip

for fossil-hunting, and for gathering shells and other collectibles to take home to display in cabinets or to use for creating garden grottos. The celebrated fossil-hunter Mary Anning was at work on the cliffs of Dorset, and George Eliot and her partner George Lewes were busy gathering material for his *Sea-side Studies*. *Punch* published a satire of this new passion: a cartoon depicted female amateur scientists in their crinolines, looking much like the sea anemones they were hunting. The natural fauna and flora of the shoreline was dried, pressed, bottled and encased in glass to adorn drawing rooms and conservatories. The shore had become a safe space for women and small children to conduct their own exciting explorations, while men went to the Arctic and Africa.

The passion for beachcombing was disastrous for the ecology of the English shore. The poet Edmund Gosse, son of the naturalist Philip Gosse, lamented that while the coast had been a treasure trove in his early childhood, by the end of the century, 'all that was over and done with. The ring of living beauty drawn about our shores was a very thin and fragile one. It had existed all these centuries solely in consequence of the indifference, the blissful ignorance of man . . . No one will see again on the shore of England what I saw in my early childhood.' The educational dimension of the English seaside experience is still evident in many resorts in the form of tourist attractions such as aquariums, zoos, and tours to birdwatch and spot seals, as well as the seafront trade in shells and fossils. Dyce's painting was firmly in the Romantic tradition – as so many subsequent artists' interpretations have been – viewing the shore as an arena for existential angst, prompting philosophical questions about the meaning and purpose of human life in the cosmos.

It has become a cliché. A 1988 Nescafé advert portrayed the liminal role of the seaside as a place of new beginnings: a young woman parks her VW Beetle just as first light dawns on the sea. Her cheeks are streaked with tears, the car's radio is set to the shipping forecast (that much-loved reminder of the British island nature). She heats a mug of water with a filament powered by the cigarette lighter as the sun rises and she gets out of the car to face the sea, a broad smile over the soundtrack: now she can see a way forward and it is going to be a lovely sunny day. The strapline, 'Make a Fresh Start', represented a new image of female independence and self-reliance. In my twenties it struck a chord, and I would find myself humming the soundtrack; I dreamt of owning my own VW Beetle and even sipping Nescafé alone as the sun rose over an empty beach. The advert brilliantly played on those cultural touchpoints of our relationship with the seaside as a place of existential reckoning, a distant echo of Dyce's Pegwell Bay. Only in this interpretation, the beach is a place of epiphany and renewal.

The centre of Ramsgate has a striking self-importance quite distinct from its Thanet neighbours; it has a clock which sets its own time, predating Greenwich Mean Time; Ramsgate Mean Time is 5 minutes, 41 seconds ahead. In 1821 Ramsgate became the only harbour in the UK to receive a royal designation, after a visit by George IV, and an obelisk was erected in gratitude, based on a reproduction of Cleopatra's Needle. The town has a history of thinking big. Samuel Taylor Coleridge, Jane Austen and Wilkie Collins were amongst its many famous visitors, and the future Queen Victoria convalesced here for six months as a teenager, two years before she became queen. One of her neighbours (and a friend) was the Jewish financier and philanthropist Sir Moses Montefiore, who had bought an estate in 1831 on the east cliff of the town, where he subsequently built a synagogue and mausoleum. Meanwhile, on the west side of town, the architect Augustus Pugin built a home and a church, and his son followed suit by building a monastery. One a Jew and the other a Catholic, they represented Victorian England's new multiculturalism. Montefiore, passionately both Jewish and English, became the first President of the Board of Deputies of British Jews, a post he held for thirty-nine years. Across town, in his bay window overlooking the sea, Pugin designed a national architecture for the great institutions of state – the Palace of Westminster and Big Ben – and campaigned for a return to the artistic and spiritual values of the Middle Ages.

A century later, in the 1930s, my father arrived at Ramsgate Abbey School, housed in buildings designed by Pugin, as an eight-year-old boy from Finchley in north London. The hammer-beam ceilings arched over the classroom, hallways were elaborately decorated with wall paintings and tiles, while the chapel and the library's shelving were detailed with Gothic motifs. The impact on my father of Pugin's neo-Gothic and love of medieval craft was lasting: this schoolboy absorbed the rich grandiosity of the whole endeavour, and, at ten years old, was precociously commenting on Pugin-designed candlesticks in his letters home, and had developed a passion for woodcarving. He later became a sculptor. Amongst his papers after his death were a sheaf of photos of the school in Ramsgate: the neat rows of small boys with their identically clipped hair, pale faces, knobbly knees and striped ties, forming a geometric pattern of dark and light; all individuality stripped out – no smiles, their arms, hands and legs identically positioned. The Pugin interiors were impressive but austere; an environment designed to intimidate and institutionalise. By an ironic twist, the monastery has now been taken over by an Indian Catholic monastic order of a particularly fervent form, another chapter in Ramsgate's history of attracting missionaries. The most famous, Saint Augustine, landed on the Isle of Thanet in 597. Known as the father of English Christianity, he had been sent by the Pope from Rome on a mission to convert the heathen island.

The once self-confident Ramsgate, brimming with huge visions of nation, identity and faith, is now full of uncertainty. Pugin's bay window overlooks the port, but the ferry no longer crosses to France, and it is not clear what future the warehouses, cranes and docks have. At Ramsgate Sands, once the bustling beach of the Victorian painting, the site of the former Pleasurama amusement park lay derelict for twenty years after a fire. Work has finally begun on a development of luxury flats when I pass by, with prices up to £750,000. Some Ramsgate locals look nervously at the experience of their near neighbour, Margate, and its property boom. Rebecca Gordon-Nesbitt was the Labour Party parliamentary candidate in the 2019 general election (and lost in this traditionally Conservative stronghold); an academic from London, she has lived in Ramsgate for eight years. Her research background has included health inequality and arts-led regeneration, and she takes a dim view of the property developments of recent years in Thanet.

'They are owned by speculators, who make profits which never touch the area, and we need money that stays here. I don't see how the luxury marina development proposals for Ramsgate harbour will help the community; there just hasn't been enough public investment. Margate seems to have measured its success in the rise of property values – there was a frenzy and it was a gold rush for estate agents – but housing needs to be built for locals, not DFLs. The area suffers from low educational levels, health inequalities and lack of work, because there is no viable economy. In Margate's Cliftonville, hundreds of people are dependent on food banks, and that's not just because of Covid. I saw what happened in Glasgow [where she used to live] with the arts-led regeneration; it just exacerbated the inequality and the centre of the city became very expensive. Thanet has similar challenges, albeit on a different scale. It needs a further-education college, better schools, and a viable economy.'

Gordon-Nesbitt invites me to a town-hall meeting on Zoom to hear the latest discussions about the future of the town, and I join postage-stamp-sized images of Ramsgate citizens on their sofas with cups of tea or glasses of wine. Proceedings start with a twenty-five-minute introduction from an academic living locally on the shortcomings of the 'extractive capitalist system', the horror of luxury housing, the pesky DFLs, and a plea to view the possibility of 'anti-capitalist revolution' with an open mind. One by one, Ramsgate protests from the sofa, impatient with the theory, and keen to get down to the really important issues: the dust the harbour produces, for example, and the shifting sands and tidal flows of the North Sea and the Channel, which could clog up the sea approach to the port; a Green Party member warns us that the port could be under water by 2050, given the climate emergency.

'People have lost hope of anything ever changing in Thanet. Everything feels like a battle here,' lamented one resident later.

During her two years of campaigning for the general election, Gordon-Nesbitt saw how deeply divided the community was. The Brexit vote here was high – 64 per cent in Thanet – and older voters frequently told her, 'We've gone downhill, it was better before we joined the EU.' She saw how people 'had no hope, no prospects, no security. They broke down in tears about their battles with the benefits system. It breaks your heart to hear.'

Jacqui Ansell, an art historian, has lived and raised a family in Ramsgate for twenty-three years and loves the town. She and her husband originated in other parts of Kent, and she is familiar with how the rest of the county refers to 'East Kentitis', which she defines as 'a sense of hopelessness and pessimism'.

'Thanet may no longer be an island in reality, but it is in the imagination. English history started here – Julius Caesar landed near Deal, St Augustine arrived as a missionary. Thanet is hard-wired to repel invaders. We can see France – it's nearer to us than London – and it sits there, glinting on the horizon, either threatening or tantalising.'

She points to another aspect, often overlooked, of Ramsgate's distinguished history: the town's motto is *'Perfugium miseris'*, 'Refuge for those in need', and is carved into the Georgian light-house at the end of the western harbour arm – a reference to the protection the harbour provided for shipping. The harbour was rebuilt after a devastating storm in 1703, which is believed to be the only hurricane to have hit British shores, when thousands of lives were lost. A recent art installation emblazoned the motto in big plastic letters on the harbour wall as a reminder of this history.

Ansell is acutely aware of the deep divisions in the town and saw how Brexit exacerbated the polarisation. 'DFLs and locals go to different pubs, cafes and restaurants. I am trying to find a way to bridge the divide. We have a very strong community ethos, with lots of amazing volunteers, and the creatives have started all kinds of local projects, often on low pay. To people born and brought up here, the rich seaside-resort heritage is unimportant; they want fishing, and they want nearby Manston Airport to survive. They want local jobs,' she says, adding that even in the schools the DFLs and the locals tend not to mix, since Kent has retained the divisive grammar-school system.

'The only part of their history which is vivid is the Second World War, and there are various re-enactment events, with dressing up and references to the Dunkirk spirit. In the war Ramsgate was heavily bombed [on 24 August 1940, 500 German bombs fell on the town in five minutes], and three hundred families took to living in disused tunnels as troglodytes. The subterranean community developed its own canteen, hospital, shops and barber. But much of the rest of their history fails to resonate.

'People from some of the poorest estates don't even go to the beach often, because the bus fare for a family is too expensive. It was the same in Folkestone – I took a group of school children round the National Gallery once, and some of them had never been to the beach.'

The local issue emblematic of these deep divisions is the campaign to save Manston Airport, which closed in 2014, leading to the loss of 150 jobs. Formerly an RAF airstrip, its association with the Second World War triggered a strong campaign to reopen it, and a crowdfunding appeal raised £85,000 to fight a planning battle in its favour. The government stepped in to grant permission (later successfully challenged in the High Court by a Ramsgate resident), but a second independent report repeated earlier conclusions that it was not viable as a commercial freight hub. In August 2022 the government again stepped in to grant permission and the developer hailed the decision as 'breathing new life into an historic aviation asset'. But within a month another legal challenge was lodged with backing from Ramsgate town council. The issue seems to be caught in a loop and it has bitterly divided the community along comparable lines to Brexit, and was crudely summarised as local jobs vs DFLs preoccupied with climate change.

Ansell admits she was frustrated: 'I don't have a local job, local jobs don't exist – that's the way the world is. I talk about reinventing the town in terms of heritage and tourism. There were suggestions that the development of Manston Airport would provide 20,000 jobs; three years later that figure fell to fifty. People have been infantilised and are demoralised and depressed, they doff their cap to the Tory MP who stokes divisions and blames DFLs.'

In late 2020 Manston Airport had reached the national news when it was pressed into national service at a time of crisis, only this time it was not for flying planes, but as a lorry park to ease congestion at Dover provoked by Brexit. Aerial shots showed scores of lorries parked on the runway. The symbolism was richly ironic: an airport associated with fighting for freedom from a tyrannical European power, Germany, was now in service again in a cause many saw as analogous – escaping the tyranny of the European Union. In this ersatz struggle for 'freedom', the airport's purpose was banal: a lorry park.

By October 2022, Manston Airport found itself centrestage of another national border crisis; it had been designated as a short-term processing centre for new migrants arriving on the Kent coast but within a few months it was desperately over-crowded. A facility designed for 1,000 was holding four times that number, and many for much longer than the statutory five days. The chief inspector of borders and immigration declared himself speechless at the 'wretched' conditions, and the government was forced into hurried – and shambolic – measures to move people. Manston Airport had become the emblematic site of how a strand of English nationalism was fixated on its borders at the price of dysfunctional and shocking misgovernment.

The issue which most disturbed Gordon-Nesbitt during her campaigning was the deep anxiety about migrants. 'This fear of invasion is in the Thanet psyche, and politicians have exploited it. Nigel Farage tried to win the seat in 2015. If you dip into some of the Facebook groups, there is no end of abuse of Travellers, and a deep anxiety about being cornered, and cut off – we are surrounded on three sides by water. It's in the DNA.'

*

The light is already thickening by early afternoon on this November day as I cycle along the main road through the centre of Ramsgate, past the grandeur of its late-nineteenth-century brick balustrades and promenades, and up Madeira Walk, with its dramatic layout of rocky outcrops and palms – as if a chunk of the Azores had landed in Kent. I am heading back to Margate and stop for a swim en route in Botany Bay, one of several small, picturesque coves surrounded by chalk cliffs. By 3.30 p.m. the air is thick with salt and encroaching darkness, but there are still a few lockdown pilgrims perambulating. As I strip off and venture into the meek, chilly waves, lights on the ships on the horizon are blinking. On one, powerful spotlights trained on the ship's deck glow brilliantly in the gloom.

For several years I have swum on the beaches along the Kent coast, on day trips to places such as Whitstable, Herne Bay and Reculver. I thought it was the closest and cleanest seawater to my home in east London, but I was wrong. The Foreness pumping station, which I passed just east of Margate, is part of the infrastructure run by Southern Water which regularly releases tons of raw sewage on this coast. It's a practice that takes place across the UK, evident even in picturesque Runswick and busy resorts such as Scarborough and the well-heeled towns of Hampshire; storm overflows, designed for emergencies such as extreme weather, are routinely used by water companies to spill untreated sewage into the sea. The worst culprit has been Southern Water, and in 2020 it was fined a record-breaking £90 million by the Environment Agency for its history of breaking regulations. It pleaded guilty to fifty-one offences on the Kent and Sussex coasts between 2010 and 2015. The judge described it as 'the worst case brought by the Environment Agency in its history', and estimated that between 16 and 21 billion litres of untreated sewage had been poured into the sea. Even such a huge fine was feared insufficient to end the practice or galvanise companies into the massive cost of improving the infrastructure. In October 2021 five Thanet beaches had to be closed, including Botany Bay, and in June 2021 eleven beaches, including Margate, were closed after the Foreness pumping station was damaged by lightning. Furious residents of Whitstable and Margate launched a ratepayers' revolt, and won the backing of an unlikely alliance of Bob Geldof (who has a home nearby) and local Conservative politicians. In October 2021 Conservatives voted down an amendment which would have placed a legal duty on water companies not to pump sewage into rivers and seas, provoking outrage; Tory MPs – many with seats in seaside towns – defended their action by arguing that such a duty would require an exorbitant investment in the Victorian infrastructure. I haven't been back in Kent waters since.

Lockdown has eased and it is Folkestone in midsummer. The sea is an oily calm, shifting stealthily, as if under a heavy blanket. France is a line of graphite on the horizon. My eye is caught by a colourful hut perched on the edge of the harbour, as if about to fall on to the muddy sand; another, in candyfloss pink, is in the middle, and later, when the tide comes in, it appears to float.

Folkestone has the biggest collection of public art in the country, part of its bid to reinvent itself after its decline as a resort and long history as a port. With the opening of the Channel Tunnel in 1994, the ferries came to a halt and it fell on hard times. Pockets of deep poverty persist, but with the help of local businessman Roger de Haan, the former owner of the Saga holiday company, huge investment in the town has led to regeneration. The narrow winding streets of the Victorian centre have been renovated as a Creative Quarter, with art galleries, gift shops and cafes, and the Harbour Arm has been converted into studios and bars. Work is underway on a new development of flats on the beach, within barely a hundred metres of the breaking waves.

The former railway line which used to carry passengers to the waiting ferries cuts through the town down to the quayside. For a century, Folkestone's railway station was the first or last touch of England. The handsome station has been part of the town's renovation project: its elegantly curved platforms with their benches and waiting rooms have been beautifully restored; the rails are still evident in the ground, but now flower beds run between them, planted with valerian, sea cabbage and thrift. The signs 'SORTIE' and 'WAY OUT' swing gently in the breeze; once, waiting passengers would have been sheltered from the Channel's wind and rain under the wooden canopy. The restoration's attention to detail is such that one can almost hear the slamming of train doors, the hoot of steam engines, the screech of brakes and porters' cries.

Of the millions of journeys the station witnessed in its working life, those which come most vividly to mind are the ones taken by soldiers heading to the trenches in the First World War. During the four years 1914–18, a total of five million travelled through the two Kent ports of Dover and Folkestone, and 1.5 million of the injured returned on their way to hospitals and home. Folkestone seems haunted still by this catastrophe, and the visitor can't avoid the ghosts. Remembrance Road runs up the hill to the promenade, where a steel arch, nine metres high, commemorates the role Folkestone played in both wars. On the pavement below is a dedication: 'In undying memory of the many million officers and other ranks, both men and women, forming the Naval, Military, Air and Red Cross services of the King's Imperial and Colonial Forces, who crossed the seas 1914–1919 to defend the Freedom of the World.' Underneath is a phrase in brackets: '(Dedication taken from the Harbour Canteen books)'.

The dated wording made this history raw and literal. The railings are hung with hundreds of handknitted poppies, a community project to mark the recent anniversary of VE day. A little further along the Leas, the clifftop promenade, I stop by a bench and am startled by a repetitive, disembodied voice: 'March 2nd, March 2nd, March 2nd.' The concealed recording has jammed. I move on to the next bench and another voice begins: an upper-class young woman reading from a letter. 'June 1915. How goes the shooting? Will your company get lots of Huns?' Her chatter turned to the frocks she was planning for her trousseau, and the ring that her fiancé had given her, and how she had enjoyed the cuddles during their weekend together. 'We will motor to Folkestone,' she wrote, 'following the route you took on your march via Maidstone.' She signed off as Marjorie. Her cheerful missive drops into an empty silence: did her exuberance survive the following months and years; did her fiancé survive?

On the quay where the soldiers would have lined up to board the ships stands Antony Gormley's cast-iron statue of a naked man, part of his installation *Another Time XVIII*, a series of 100 cast-iron figures dispersed around the world (including two in Folkestone, and one in Margate). The stone steps which lead down to the embarkation point are covered in green seaweed, left behind by the high tides which submerge the statue. He looks out towards the White Cliffs of Dover, alert and hesitant, as if he is about to jump off his seaweed-covered perch. Is he about to escape, or is he a sentinel guarding the English coastline? Unlike the monuments evident elsewhere in the town, he is hidden, waiting to be discovered, and his statement is a whisper: what if?

Folkestone was once a fashionable Edwardian resort with magnificent hotels overlooking the sea. The cliffs were landscaped with gardens, grottos, caves, rockeries and winding paths in imitation of the rocky coastline of Sorrento. Look closely, and the concrete and brick are evident beneath the weathered cement, but the overall effect is ingenious and, looking down the steep slope, the green sea sparkles between the fir trees, flowers tumble over the walls and beds brim with agapanthus and verbena. It is much loved, and the grottos and caves are decked with memorials – photos, dead flowers and the occasional candle.

The beach at Sunny Sands has narrowed to a small strip,

provoking chaos as beachgoers scramble to rescue their bags from the encroaching tide. I slip in quickly in this maelstrom of parents and children screaming with delight as the waves come ever closer. Once I'm away from the shore, the noise recedes, all the hubbub briefly distant. My piles of clothes now threatened, I race the waves back. Cornelia Parker's statue of a mermaid watches benignly, sitting modestly beside the beach, the only bather unperturbed by the high tide.

By the harbour, my crab sandwich attracts the attention of a plump seagull, which flies so close it almost knocks off my sunglasses. It settles a few feet away, a beady pink eye fixed on my lunch. It keeps inching closer until, alarmed at the cat-sized predator, I flee, and it waddles after me. Eventually it gives up, turning its attentions to another victim.

In September 2020 the Home Office took over a dilapidated former army barracks as a short-term hostel for 400 asylum seekers, mostly young men. The story of Napier Barracks unfolded like a slow-motion car crash, attracting national and international condemnation. Napier Barracks had been unused for several years pending redevelopment, but the Home Office maintained that it was needed to house asylum seekers to meet a 'bulge in the system'. The decision inflamed a toxic combination of local sensitivities: the recent rise of migrants crossing the Channel, respect for military heritage, and fear of Covid-19. It turned ugly, and far-right activists filmed residents of the temporary hostel from the pavement for their online publicity; sheets had to be put up along the chain-link fence to shield the asylum seekers. 'Little Veteran' posted regularly on YouTube, attracting thousands of views for his coverage of ambulances arriving at Napier Barracks; 'How much have these migrants put into the NHS?' he asked over footage of Second World War history and war memorials, before commenting on the impact of post-traumatic-stress disorder on veterans. One hostel resident told the local news website KentLive, 'When we get out of the camp, I see people looking at us as if we were a herd that came out of a cage. Some cross the road when they see us walking on the same side, and some look at us with disgust.' Residents in the neighbouring streets complained. 'We're a very small island and we're very full, and whilst I'm happy to help anyone in genuine need, a lot of the people coming here are economic migrants. I don't agree with that at all, and there's also the threat of terrorism,' said one. Another added, 'I understand that they've come from bad places, but what about local people who are hard done by? To put foreign nationals in an army barracks and give them health care and food is a disgrace when we've got homeless veterans on the street. We're also in a pandemic. How's it fair to increase the health-care demands on local services?'

The British Red Cross declared that the Napier Barracks were not appropriate for an asylum processing centre, going against its usual reluctance to criticise government decisions. Refugee charities raised concerns that the facilities were inadequate and could trigger post-traumatic stress in asylum seekers with experience of detention and torture. In January 2021 numbers had to be reduced because of a large outbreak of Covid-19 in the overcrowded barracks. Urgent complaints were lodged with the UN special rapporteur on the rights of migrants, and a report by government inspectors criticised the Home Office, concluding that conditions in the barracks were cramped, filthy and unsuitable for inhabitation, and that many of the residents had mental-health issues. In June 2021 the High Court ruled that the barracks failed to meet minimum standards, but the Home Office pressed on, and announced that it intended to continue using the barracks until 2026. Priti Patel, the then home secretary, insisted that 'it was an insult' to suggest that buildings that had housed 'our brave soldiers' were not good enough for 'these individuals'. In June 2022 a judge ruled that the Home Office had breached the Equality Act and that Napier's use was 'unlawful'. In October 2022, despite inspections ruling that the place was still unsuitable, it was housing hundreds of asylum seekers. Bravely, some locals mounted counter-protests, waving placards welcoming the migrants. The long-running story over the months of national lockdowns led to deep divisions in the town. Meanwhile, the number of migrants desperate enough to attempt the Channel crossing continued to rise, breaking new records in the summer of 2022, despite the use of the Royal Navy and Priti Patel's draconian decision to launch a scheme to send asylum seekers to Rwanda. Neither proved a deterrent, and critics argued that, without legal routes to asylum in the UK, migrants would continue to risk dangerous journeys in small boats to reach the Kent coast.

Folkestone and Dover have been the gateway to England for centuries, and as the traffic of people and goods has accelerated, so has the insatiable demand for access – bigger motorways, tunnel entrances, lorry parks, service stations – all shoehorned into a small area of steep chalk downland. The result has been some spectacular engineering, so that roads cut through hillsides and plunge into gaping tunnels bored into the chalk. A stream of cars and lorries are swallowed up and spewed out on to Dover's docks, while the Channel Tunnel entails a skirt of tall metal fences and concrete sidings before disappearing underground on its way to France. The once picturesque landscape has been brutally sacrificed to national need; sheep graze the grass and gorse blooms on slopes of downland which amount to traffic islands. The A20 to Dover shoots past the steep Shakespeare Cliff, which looks as if it has lost half of itself, sheared away by hungry tides, its chalk edge still raw. Walkers on the coastal path are deafened by the roar of traffic. The White Cliffs of Dover, one of the most famous landscapes in the country – the image of Second World War propaganda and the subject of Vera Lynn's famous song – has been hedged by roads bearing millions of vehicles on their way elsewhere.

At the base of Shakespeare Cliff lies Samphire Hoe, a new country park made from the spoil produced by the construction of the Channel Tunnel. Five million tons of chalk were dug out of the seabed and deposited here to form undulating hills, with grazing cows, reed-fringed ponds, heathland grasses and wild flowers, between the cliffs and the sea. When I visit, fishermen line the concrete sea wall, and families are picnicking amongst the meadow grasses. Samphire Hoe has a peaceful innocence in an area where the past threat of invasion is evident at every turn.

On the outskirts of Folkestone stands the Battle of Britain memorial, dedicated to the young pilots who managed to hold a German invasion at bay in 1940. In the centre of Dover, a memorial commemorates the 1940 evacuation of the British army from the beaches of Dunkirk, assisted by volunteers in an armada of boats. These stories of solitary defiance have become the most precious of national histories. The memorial which stands on Dover's seafront is a rust-stained section of armoured plating taken from a Second World War German long-range gun based twenty-one miles away on the cliffs of Pas-de-Calais. A pale grey eagle perched on top of a bomb was painted by the German troops, and below, they marked the daily tally of bombs fired at Dover between 1940 and 1945. Few other British towns experienced such continuous and intense bombardment. The worst came in September 1944, when the Germans raced to use up their stock of shells before they were overtaken by the Allies.

If the land around Dover has been cut to ribbons, the town itself has been disembowelled. The A20 thunders through it, spawning ungainly shopping outlets which dwarf the few nineteenth-century streets surviving from an earlier era. A shabby underpass connects the town centre and the seafront, and a young man was begging from the few passers-by. Dover was once a resort as well as a town, and some of the terraced houses still bear traces of their original elegance, while others have been converted into shops (inevitably, a large number are charity shops). A prominent tattoo parlour was emblazoned with the motto, 'No pain, No gain.' The town was festooned with Union Jacks and St George Cross flags.

A recent renovation project has rescued some of old Dover, and in the middle stands a grand monument to the 'Indian

Mutiny' (sic) of 1857–9 and three campaigns at Oude, Delhi and Rohilcund, under the Latin motto *'Celer et Audax'*, 'Speed and Daring'. On the nearby seafront, it was as if the ghosts of retired colonial servants were smoking cigars on the wrought-iron balcony of the yachting club overlooking the beach. Perhaps one of these tall terraced houses was where the Victorian poet Matthew Arnold had his honeymoon, saw the moon from his window and wrote his famous poem, *Dover Beach*:

> *The sea is calm tonight.*
> *The tide is full, the moon lies fair*
> *Upon the straits; on the French coast the light*
> *Gleams and is gone; the cliffs of England stand,*
> *Glimmering and vast, out in the tranquil bay.*
> *Come to the window, sweet is the night-air!*
> *Only, from the long line of spray*
> *Where the sea meets the moon-blanched land,*
> *Listen! you hear the grating roar*
> *Of pebbles which the waves draw back, and fling,*
> *At their return, up the high strand,*
> *Begin, and cease, and then again begin,*
> *With tremulous cadence slow, and bring*
> *The eternal note of sadness in.*

Arnold's 'tranquil bay' is a thing of the past. The ships coming in and out of Dover's port are a continuous presence on the horizon. Undeterred, some families were picnicking on the beach and a large Asian family were enjoying themselves; two teenage girls launched themselves into the waves. Dover Castle loomed overhead on the cliffs above, witness to centuries of preparations to fend off possible invasions, and the launch point for British armies fighting in France. At first, it seemed incongruous that this shabby town was the setting for Arnold's great poem describing the Victorian crisis of faith. Having known and loved the poem long before I visited the town, it seemed a sorry setting. The themes had a grandeur and scale quite out of proportion with the twenty-first-century version of Dover. This thought was quickly succeeded by another: that the bitterly bleak concluding lines aptly capture how Dover, as the country's most famous border town, has been the scene of multiple crises in recent years:

> *And we are here as on a darkling plain*
> *Swept with confused alarms of struggle and flight,*
> *Where ignorant armies clash by night.*

In late 2020 lorries backed up on the A20 to enter the port was a graphic illustration of the chaos of Brexit. Meanwhile, increasing numbers of migrants were arriving at Dover's docks in life jackets and emergency blankets to disembark from coastguard boats; the flimsy inflatable dinghies on which they set sail from France drifted on to nearby beaches. Here was the pinch point of geo-political tensions, and the desperate movement of people fleeing poverty and violence. When Arnold stood at the window of his hotel and admired the moonlight, was this what he was imagining of his country's future?

The migrants arriving on Kent beaches have provoked the formation of groups purporting to 'monitor' the coast. One has claimed the title of 'South-East Coastal Defence'. The far-right former UKIP leader Nigel Farage took up the issue in 2020, posting videos on YouTube of heavily laden dinghies under the caption, 'If people tell you the invasion isn't happening, show them this footage.' His inflammatory commentary talked of a 'massive criminal' operation making 'vast sums of money'. The then prime minister, Boris Johnson, and Home Secretary Priti Patel tried to outmanoeuvre Farage. In September 2022 Patel's successor at the Home Office, Suella Braverman, made the stopping of small boat Channel crossings the key target for her department after another record-breaking summer saw 8,747 crossings in August alone. Border enforcement now involves drones, helicopters, night-vision equipment, naval patrols, amplifying the sense of threat without reducing the numbers attempting the crossing. On 30 October 2022, a man travelled from High Wycombe, Buckinghamshire, to detonate incendiary devices at a centre for processing migrants in Dover before killing himself; police designated it a terrorist incident motivated by right-wing extremism. After a long history of being on the frontline of conflict, the threat Dover now faces is not from over the sea. In response, local refugee groups have initiated their own patrols to offer support to new arrivals; they argue that the legal channels for claiming asylum have been so tightly restricted that people have no option but to turn to the danger of a Channel crossing. UK figures for granting asylum remain very low in comparison to many other countries in Europe, such as Italy, Greece and Germany. In 2020–21 new legislation was introduced in parliament proposing the even more draconian step of making illegal migration a criminal offence. The coast around Dover is turning into a form of militarised zone, and its targets are human beings desperate for a better future.

Up on the White Cliffs above Dover, the National Trust cafe offers a bird's-eye view of the town's docks. Seventeen per cent of British trade passes through this port, and the carefully coordinated movements of boats and vehicles below was mesmerising, as ferries unloaded and loaded several times an hour. The use of time and space was precise, there was no slack in this system, no spare parking bays or queueing space. There was no illustration more vivid of Britain's tight economic links with its biggest trading partner, the European Union; but it seemed that some of the most ardent Brexiteer ministers had never had the opportunity to stand on the White Cliffs and grasp the concentration of British–EU trade through this port – as they admitted – and had not foreseen the chaos of late 2020 at the port.

The sun had broken through, and the paths over the White Cliffs were busy with visitors. They arranged hair and smiles to pose against the iconic backdrop of chalk and sea for selfies. A group of four elegant young Muslim women, dressed in long black skirts, jean jackets and neatly tied headscarves, were laughing excitedly as they photographed themselves. 'It's the first time we've been,' they explained. Incongruously, two ornate velvet armchairs and a sofa had been arranged on the cliff edge for a photo shoot; the team was hard at work, trying to secure white tablecloths and flower displays, and styling the hair of two black models, dressed as a bride and bridegroom.

The White Cliffs of Dover was invented as a national symbol during the Second World War, reflecting the role of Kent and its ports in both world wars. Much of its fame is down to the song 'The White Cliffs of Dover', first made famous by Vera Lynn in 1942:

> *There'll be bluebirds over*
> *The white cliffs of Dover*
> *Tomorrow, just you wait and see.*

> *There'll be love and laughter*
> *And peace ever after.*
> *Tomorrow, when the world is free*

The shepherd will tend his sheep.
The valley will bloom again.
And Jimmy will go to sleep
In his own little room again.

It's been called an alternative national anthem, and a new version by the ninety-two-year-old Vera Lynn reached number one in 2009. Its nostalgic pull is strong, yet ironically the lyrics were written by an American who had never been to Britain, at a time when shepherds were already a rarity around Dover, and the bluebird is an American bird which has never been seen flying over any part of the UK. But it reinforced the symbolic significance of this corner of south-east England. As the empire was dismantled and Britain became a member of the European Union, and the Channel Tunnel was built, Dover became the umbilical cord to the continent and this new diplomatic alignment, marking a shift away from the Atlantic-orientated ports of Liverpool, Glasgow, Southampton and Portsmouth. Perhaps inevitably, the symbolism of the White Cliffs was repeatedly invoked in the bitter debate about Britain's relationship with Europe. In 2015 Farage's UK Independence Party launched an anti-immigration campaign at the White Cliffs, using an image of an escalator up the cliffs crowded with migrants. In 2017 a massive cut-out of the then prime minister Theresa May was suspended from a crane on the clifftop, dressed in a Union Jack and giving a V sign; no one accepted responsibility. In 2017 Vera Lynn's image was projected on to the cliffs to mark her hundredth birthday, and images of Nigel Farage, David Beckham, beer adverts and refugee charities all followed. When the UK left the European Union, several media organisations projected goodbye messages, visible only from sea at night, but instantly spread on social media by an argumentative, uneasy nation.

5

Brighton to Bognor

Sussex

On summer bank holidays the trains out of Victoria and Clapham are packed, and the traffic on the M23 is thick, as the capital moves south to the coast. Those arriving in Brighton by train are swept along in the crowds which pour down Queen's Road, drawn on by the glimpse of blue sea before they spill on to the beach. The steeply shelving pebbles may not offer much comfort, but ahead there is nothing but water. This is where London goes to meet the edge; this is the end of England. The Channel widens here and, unlike at Dover and Folkestone, one is free of a neighbour peering over a metaphorical fence. Brighton has always offered Londoners a sense of escape, freedom and adventure. The Irish novelist Anne Enright observes in her novel *The Gathering*, 'And there it is; the open tang, the calling, the smell of the sea. Such a miracle, at the end of the Brighton line, with the town stacked behind me, and behind that all the weight of England in her smoke and light, jammed to a halt here, just here, by the wide smell of the sea.'

Piers, promenades, grand terraces, squares, gardens and bowling greens: Scarborough may have invented sea bathing, but it
was Brighton which invented a new coastal townscape (with
competition from its closest rival, Margate). Brighton pioneered
a novel urban architecture designed to delight, and created
the seafront as a new space in which to parade, socialise and
sit to gaze at the sea, in ornate wrought-iron shelters. By 1833
three miles of seafront had been constructed – matched only
by St Petersburg; by 1823 it had built one of the first piers, the
iconic feature of the English seaside, symbolically bridging the
meeting point of land and water and bringing a new intimacy to
the mystery of the sea. Over the following century every self-
respecting resort followed suit, although Brighton's pier was by
far the most ambitious. Visitors could walk along it, glimpsing
the sea's movement between the planks, and stand at its head,
almost surrounded by water. Or they could look back and see
land from an entirely new perspective. The pier was the first
invention in what became a long seaside tradition of offering
new vantage points, such as towers and Ferris wheels, from
which to view the familiar world. To intensify the dizzying
disorientation, attractions were later developed for speed and to
disrupt gravity, such as roller coasters and big dippers.

Ever since Brighton's discovery by the Prince Regent in the
early nineteenth century, it has astutely managed, chameleon-
like, to meet a wide variety of tastes, offering oyster restaurants,
antique shops in the Lanes, smart crescents and grand hotels, as
well as dog and horse racing, seedy pubs and shabby amusement
arcades. Historian John Walton describes Brighton as being 'a
carnival of strange juxtapositions between fashionable high society and its imitators, and an exotic medley of Cockney trippers'.
In the early 1900s the novelist Arnold Bennett was awestruck
by the contrast with his home in the industrial Potteries in

Staffordshire, and included lavish descriptions in his trilogy *The Clayhanger Family*, parts of which he wrote while staying in the town:

> As for Brighton, it corresponded with no dream. It was vaster than any imagining of it ... he had not conceived what wealth would do when it organised itself for the purposes of distraction ... suddenly he saw Brighton in its autumnal pride beginning one of its fine weekends, and he had to admit that the number of rich and idle people in the world surpassed his provincial notions. For miles westward and eastwards against a formidable background of high yellow and brown architecture, persons the luxuriousness of any one of whom would have drawn remarks in Bursley [the fictionalised town of Burslem in Staffordshire where he attended school] walked or drove or rode in the thronging multitudes ... the air was full of the consciousness of being correct and successful.

One of Bennett's characters concludes: 'this then was Brighton. That which had been a postmark became suddenly a reality.'

In time the Edwardian glamour faded and by the 1930s Brighton was better known for crime, violence and disreputable sexual trysts. It became notorious in 1934 after the lurid Trunk Murders attracted widespread newspaper coverage, with gory details of how a man murdered his wife, cut up her body and sent the parts by parcel to different places in the country. In another celebrated case in 1936, a Hoxton mob of thirty attacked a bookmaker. But Brighton has always attracted passionate advocates, and in 1939 the artist John Piper took up the town's cause. To the astonishment of many of his peers, Piper challenged the perception of a tawdry town of cheap delights and praised 'the white façades of housing facing out to sea, the piers, the fishing boats all keeping up the seaside spirit. They make thousands of people remember Brighton.' He celebrated how the architecture of bandstands and shelters had absorbed foreign influences and resulted in an idiosyncratic style all its own. In his illustrated book *Brighton Aquatints*, he claimed that 'it is not simply the sea air that is such fun and such a change, it is the whole gamut of heightened contrasts that the seaside provides, and is strong enough to make life there seem fuller and gayer'. Piper identified a 'coastal gaiety' which was 'an essential constituent of an English identity and an English style', argued one of his reviewers. As Piper had acknowledged, nowhere typified this for the nation more clearly than Brighton. The town (a city since 2001) has given full rein to the possibilities of the seaside as a place of experiment, spectacle, performance, play, transgression and indulgence. It has walked the line between 'civilised constraint and liberated hedonism', suggests Walton, adding that always, 'the spirit of carnival bubbled close to the surface'.

Brighton's fortunes have ebbed and flowed over the 250 years of its history as a seaside resort, but for the last twenty-five years it has been regarded as the most successful in the country, and praised for its reinvention since the 1980s. Many of the factors contributing to its current vitality are difficult for other resorts to match: proximity to London and its overheated housing market, a major airport at Gatwick, and a fast railway link are crucial, as are two universities, with their population of 35,000 students. That pool of graduate talent has been key; Brighton has one of the highest proportions of adults with a degree in the country (50 per cent, compared with a national average of 38 per cent), and 41 per cent of the workforce is employed in knowledge-economy jobs, including a wide range of creative industries. As property became increasingly unaffordable in London, Brighton benefited from the spillover in the 1990s; now the house prices almost match those of the capital. A brisk business in conferences attracted ten million visitors a year pre-Covid, a major contributor to the economy. Meanwhile, it retained the positives of its seaside-resort history, with a reputation for tolerance and diversity – as early as the 1920s there was an established gay and lesbian scene. One of the first Pride marches was held in Brighton in 1973, and by the 1980s it had one of the biggest gay and lesbian communities in the country, with a plethora of LGBTQ organisations. It had broken the classic seaside-resort mould of low-paid, seasonal work, and in 2016 celebrated its confidence with the launch of its 162-metre observation tower, the British Airways i360.

Brighton has come a long way since the early 1980s when I moved there from North Yorkshire, aged sixteen. For the following three years, I was intent on reinvention, and it was the perfect place to do it. My mother and I ended up in Brighton by sheer happenstance, like many people arriving in a coastal resort: a combination of a broken marriage, the need for cheap housing, and the desire for a fresh start. I joked at the time that it was as far away from my father as my mother could get without actually emigrating. My sister had arrived a year earlier to study at Brighton School of Art, and, apart from a few of her friends, we knew no one. My mother bought a small terraced house in a shabby part of Brighton, and before long the house was full of students. Some came for a meal, some for a night, and some moved in as lodgers. Then my mother began joining groups of various kinds. I never knew who or what I would find when I came home – prayer meetings, poetry readings, singing groups, my sister on the sewing machine for her textiles degree, or the small sitting room full of someone's friends. The sense of liminality was visceral: all was in flux.

A short bike ride away was the beach, and I never lost the
sense of febrile excitement that we were on holiday from our
'real' life of rural Yorkshire and my former convent boarding
school. When we first moved in, my siblings and I visited all
the attractions, rode on the dodgems and the mini train, and
played on the slot machines in the amusement arcades with
spare change. I feared that the magic of Brighton couldn't last;
it was too much fun and no one was telling us off. Its freedom
was intoxicating.

The Pavilion was inevitably one of the first sights on my
arrival. In 1815 the Prince Regent had indulged his every
whim in the construction of his ornate residence in the town,
and its gaudy implausibility as a royal palace astonished me; it
represented excess and playfulness, not power and authority. I
deduced that Brighton was a place where one could shake off
convention and let the imagination run where it might – fake
bamboo, iron palm trees, gilt pagodas and dozens of Mughal
cupolas. Experimental, flamboyant and playful was the order of
the day. Derided in its time as a 'gilded dirt pie' and a 'minaret
mushroom', the contemporary accusation is that it is a shocking
case of cultural appropriation; in his book on the seaside, Travis
Elborough accuses it of being 'an almost sacrilegious twist on
Islamic religious architecture' and likens it, in terms of cultural
insensitivity, to the pork lard used in Enfield rifles that played
a key role in provoking the Muslim soldiers in the 1857 Indian
Rebellion. He has a point; its riotous concoction of styles, Indian
and Chinese and 'a generic tropical' look – palm trees formed
the pillars holding up the high kitchen ceiling – represented
a country intoxicated in the early nineteenth century by its
imperial project of borrowing, extracting, reaping and pillaging.

But even in my excited discovery of the town in 1980, it
was evident that its glamour and fun had faded. The town had

hit hard times. Its tourist income had declined, and Brighton was struggling with a reputation as shabby and rundown. The famous quote of playwright Keith Waterhouse just about summed it up: 'Brighton is a town which always looks like it is helping police with their inquiries.' Brighton was the inspiration for the fictional seaside town of the novel and subsequent television series *The History Man*, about a radical, priapic academic, and its lead character, a sociologist, is relieved to discover that 'it's a problem town, full of hippies and dropouts. It's a town you can run to and disappear. There are empty houses. Visitors are soft touches. Lots of marginal work.' Much of that was still true when I arrived, but a new spirit had already begun to take hold, encouraged by its large and growing student population, championing tolerance, environmentalism and an idealistic desire to change and reinvent. They wanted to find a new way to do business – The Body Shop, with its advocacy of no animal testing and natural beauty products, emerged from Worthing, just down the road – and a new way to eat, in vegetarian cafes (I drank herbal teas and ate stodgy flapjacks in Food for Friends) and wholefood shops such as the co-operative-run Infinity Foods founded in 1971 (still going strong as one of the biggest organic, fair-trade wholesalers in the south-east). Brighton was a dream factory of idealists, fantasists, escapists and adventurers, and my project of reinvention was one of many I was witnessing amongst friends, neighbours and even my own family. My mother and sister were both launching themselves on dramatically new life trajectories.

I got a Saturday job on the till at Boots and, with my un-familiar wealth (£7 for a shift), I scoured the many second-hand shops with my sister and picked out all manner of extraordinary outfits. My sister bought an ice-blue 1950s taffeta cocktail dress, and a stained worker's boiler suit, and I borrowed both; we bought exquisite hand-embroidered 1940s silk blouses. They were deeply unfashionable, but only a pound a piece; now they could be museum pieces. We wore collarless granddad shirts down to our knees, and baggy jumpers. I salvaged a child's kilt as a miniskirt, and a smartly tailored pencil skirt with a neat kick pleat; every day I tried out a new look, bewildering my peers. Punk had faded, and the New Romantics and Adam Ant had made ludicrous floppy lace collars fashionable; a Boots co-worker once looked at my stained fisherman's smock and marvelled that I could wear such a thing.

My siblings and I discovered Happy Hour in the smart seafront hotels. The rooftop bar of the Metropole was our favourite: free peanuts with Dubonnet and lemonade, our go-to drink, which my sisters kindly treated me to out of their bigger earnings. We sat there in our shorts and sandals, hair still crusted with salt from a day on the beach, feeling ourselves to be daringly adventurous. We relished the anonymity of the seaside resort – so many people on the move, coming and going – and, as unlikely customers, no one took much notice of us, so word wouldn't get back to our mother. We were slowly easing off the shackles of small village life and the close, disapproving supervision of our convent boarding school. It was as if we could take great gulps of air for the first time. We shrieked with giggles on the seafront, and collapsed with laughter over anything. My father had always described my sisters' high spirits as hysteria, but in Brighton there was no one to frown and criticise.

In the middle of this intoxicating adventure, I earnestly set out to find fellow Christians, and naively stumbled into the then emergent world of Sussex-coast evangelical Christianity. As a journalist in the late 1990s, I went back to visit the house church movement I had known, and found an organisation of thousands – Sunday services were held in a cinema to accommodate the arm-waving, hymn-singing crowds. Back in the early 1980s, they were still meeting in houses in prayer groups, where the preaching went on for hours, and we were sternly advised to be wary of Satan and all his works. As girls, we should be covering our heads and be modestly dressed at all times and submit to male leadership. I became increasingly alarmed. At school, in our poky basement meeting room, where we belted out gospel music – 'Hallelujah! We love Jesus, we are saved' – accompanied by handsome boys on their guitars, one of the believers warned me that, as a Catholic, I was heading for a cliff edge, and beyond lay eternal damnation. She had a compelling duty to save me. I fled.

The flamboyance and drama of Christian evangelicalism suited Brighton. The liminality of the coastline as a place of change and a new start lent itself to the strident promises that all your sins could be redeemed and you could be 'saved' by Jesus and 'born again'. Full-immersion baptism in the sea during the summer was common. Preaching relied (and still does, judging by the Zoom services I attended in 2020 of Brighton evangelical churches) on the repetition of words such as 'new' and 'fresh', and the mounting crescendo of promises and superlatives: 'God is working anew in you, and doing something extraordinary in this community.' Depending on your point of view, it was exhausting or exhilarating. Amongst the large, transient population of a coastal town, a lot of people were looking for connection and community, and evangelical churches – or 'free churches', as they described themselves – made a point of building small groups, stressing the importance of mutual support and care.

When I came home late in the evening, sometimes my mother's Catholic charismatic prayer group was in full swing, and usually Brian, a gentle refuse collector, would be 'slain in the spirit'. This entailed a form of ecstasy, his head thrown back as he burbled incomprehensibly, the phenomenon known as 'speaking in tongues' – a 'gift of the spirit', according to the New Testament. For relief from all this intensity, I escaped to the seafront and took the undercliff path from Brighton to Rottingdean. Here, on a sea wall, I read *Dover Beach*: faith was proving hard work. Meanwhile, my mother's own reinvention was by now reaching disconcerting proportions; abandoning twenty years of being a Catholic wife and mother, she announced that we were a collective, equally responsible for cooking and cleaning, and formed a deeply romantic relationship with a man half her age.

My religious adventures reinforced the feeling that everything was in flux; like many others in Brighton, I was only passing through. My sights were set on university. The sense of instability and precariousness was written into the infrastructure of shabby Brighton, battered at that point by decades of decline. My sister's first lodgings as a student were in a grand but draughty and rickety seafront house, and, alarmingly, one stormy night, one of the neighbouring properties collapsed. In the morning the wallpaper and fireplaces of the end of the terraced house were exposed to the salt-laden winter gales, above a pile of rubble, in a manner reminiscent of the Blitz. The photograph appeared on the front page of many newspapers, a poignant symbol of decline. Nothing in Brighton felt as secure or permanent as it might appear to be, and that suspicion was reinforced by my eccentric English teacher, who had a passion for local history; he claimed that the grand Regency terraces of Brighton and Hove were shockingly badly built by speculators, and about as robust as stage scenery.

Meanwhile, the closed West Pier was a magnificent monument to decline and disintegration, and the subject of desperate battles to fund its renovation. I cycled past it every week on my way to a job in a care home in Hove, and watched the winter tides slowly dismantle its elegant timbers and iron struts. A sequence of storms and fires brought the pier repeatedly back into the national headlines. Its restoration as the most endangered Grade I listed building in the UK became a celebrated cause. Most ruins, when left to collapse, do so in private, behind hoardings. But not a pier; it crumbled in full view of the visitors packed on to the bank-holiday beach. Brighton's West Pier was perfectly silhouetted against the setting sun, and its constantly changing profile as it slowly fell apart makes it probably the most famous image of seaside decline in the country.

But not even in my fevered adolescent imagination could I have guessed at the spectacular collapse and destruction which visited Brighton twice in the time I lived there. In October 1984 an IRA bomb was detonated in the Grand Hotel when it was hosting Prime Minister Margaret Thatcher and her government for the annual Conservative Party conference. It was the closest a British prime minister came to assassination in nearly 200 years. Five were killed, including three wives of politicians, and thirty-one injured. The bomber, Patrick Magee, had planted the bomb under a bath in a bedroom five floors above Thatcher's suite a month earlier; he was convicted on the strength of a fingerprint on a registration card found in the hotel ruins.

'After Brighton, anything was possible and the British for the first time began to look differently at us,' said Magee many years later. He served thirteen years of his thirty-five-year sentence, released early under the Good Friday Peace Agreement. The *Daily Telegraph* concluded that the bombing was the 'most audacious attack on a British government since the Gunpowder Plot' and 'marked the end of an age of comparative innocence. From that day forward, all party conferences in this country have become heavily defended citadels.' Five years after the bombing, I was back in Brighton as a junior reporter for the *Guardian* at my first party conference, and, by then, the draconian sequence of security checks and passes was well established.

Three years after the Brighton bombing, the city was hit by another catastrophe. In October 1987 I was woken briefly in the night by the roar of a storm, but thought little of it; by the following morning, Brighton was littered with fallen trees, and swathes of old and much-loved woodland had been destroyed in Stanmer Park, on the town's outskirts. It was the south coast which took the full brunt of one of the most powerful storms seen in the UK for decades. The gales were funnelled down the central spine of Brighton, along the Old Steine and the park known as the Level, knocking down graceful avenues of mature trees like ninepins and transforming its appearance for several generations. My mother was devastated, weeping inconsolably. Shortly after, I moved to London; my life in Brighton had come to an end at a dramatic point in its history.

My three years in Brighton were at a time when the possibilities of how to live seemed endless. Everything was up for grabs. Was that my overheated imagination, my age, the place, or simply the times I was living through? Or some intoxicating combination of all these, shot through with a thrilling sense of unreality? My sister now confesses that she believed that we had walked into a novel; she swears that on her arrival at Brighton station in 1979, she *saw* Pinkie and Rose, the protagonists of *Brighton Rock*, in the taxi-cab rank. For my sister and me, Graham Greene's novel about a murder by a young gang leader, Pinkie, and his marriage to Rose, a waitress, to keep her quiet, with its bleak theology of damnation and mercy, was not just a masterful thriller but a form of extended sermon-cum-travel guide to the town. The biggest character of the novel is the town. Brighton comes to life on every page in vivid detail. The book was published in 1938, but, forty years on, its portrayal of the town was still accurate: the sea, the pier, the holiday crowds; the cafe where Rose could have worked, and the pubs and hotel where one might find Ida, the warm-hearted woman who spotted the danger Rose was in; the racecourse on the outskirts, and the suburbs of east Brighton creeping over the downs; the shabby streets under the railway viaduct off the Lewes Road where we lived: 'The houses which looked as if they had passed through an intensive bombardment, flapping gutters and glassless windows, an iron bedstead rusting in a front garden, the smashed and wasted ground in front, where houses had been pulled down for model flats which had never gone up.'

This shabby dereliction was well hidden, tucked behind the seafront, which Greene fondly describes: 'fresh and glittering air; the new silver paint sparkled on the piers, the cream houses ran away into the west like a pale Victorian watercolour; a race in miniature motors, a band playing, flower gardens in bloom below the front, an aeroplane advertising something for the health in pale vanishing clouds across the sky'. Later in life, Greene acknowledged that 'no city before the war, not London or Oxford, had such a hold on my affection as Brighton'.

A contemporary of Greene, the Anglo-Irish writer Elizabeth Bowen, shared his fascination for 'coastal gaiety', and in her novel *The Death of the Heart* (1938) she contrasted the etiolated emotional life of London's Regent's Park with her fictional Seale-on-Sea on the south coast, 'a riot of colour, commercialism and sexuality, where the life of sensations is unedited and raw energy is on display'. Her protagonist is staying in a bungalow called Waikiki, which faces 'the sea boldly as though daring the elements to dash it to bits'. Bowen commented that on the south coast, 'everything including the geological formation struck me as having been recently put together. And this *newness* of England, manifest in the brightness, occasionally the crudity of its colouring, had about it something of the precarious. *Would* it last?' (This was the question I had of Brighton, and the confusing answer was yes and no.) The seaside, not London, was 'the scene of fast-paced social change', Bowen added; she portrayed the inhabitants of Waikiki as loud, uninhibited, fun-loving and eager for new consumer experiences such as cars, cinema, cigarettes, cocktails, lipstick and nylons.

In the 1930s both Bowen and Greene were fascinated by a new, self-consciously modern sensibility, emerging most fully at the seaside. Greene was also ambivalent. In *Brighton Rock* the crowds on the seafront, with 'immense labour and immense patience ... extricated from the long day, the grain of pleasure'. His distaste was vivid for the poverty of empty grates, for the cheap American imported consumer habits, and the more traditional features of seaside resorts such as hurried sex, gambling, drinking and souvenirs. The novel's main protagonist, gang-member Pinkie, describes himself as 'the real Brighton': 'as if his single heart contained all the cheap amusements, the Pullman cars, the unloving weekends in gaudy hotels, and the sadness after coition'. The violence is always close at hand, lurking in the shadows, in the dripping wet under the pier, and on the shore, where 'a gull swooped down screaming to a dead crab beaten and broken against the iron foundations of the pier'.

On almost every page Greene references the close presence of the Channel and its vast body of moving water: 'Somewhere in the channel, a boat sounded a siren and another answered, like dogs at night waking each other.' The book culminates on the vertiginous cliffs of Beachy Head, the place for Pinkie's suicide, where 'hundreds of feet below the pale green sea washed into the scarred and shabby side of England'.

While I lived in Brighton, every skinny boy could be a Pinkie, every sweet-faced girl a possible Rose. Like both of them, I was an outsider, staring at the illuminated foyers of the fancy hotels, glimpsing the flurry of guests in their evening clothes coming and going, dining in the smart restaurants. Living in Brighton, you were always on the edge of other peoples' parties. Amongst my friends we swapped tips on the best hotel jobs, but my efforts to blag my way into a job as a silver-service waitress (training essential in those days) were never sufficiently convincing, and I remained in the care home, with the incontinence pads and cups of tea. As I cycled home along the seafront from a late shift, it always seemed possible I might catch sight of Ida on the arm of a new man, or Pinkie, his collar up, furtive in the shadows. Meanwhile, the sea, fluid and black, was only a few metres away, shifting uneasily on a calm night. Or on a winter evening the fierce wind slammed pub doors, rattled the pier struts, sent flags flapping wildly, and smeared windows with salt. Emerging from the warmth of a pub on the Lanes on to the seafront, you were met by the roar of the sea chewing at the beach, and waves flinging spray and pebbles at the passing cars. The seafront was always battling against decay, the painted railings streaked with rust, the glass of the shelters repeatedly smashed. There was always a largeness of life and heartfelt sense of humanity at this seaside. As Greene's Ida said to the policeman, 'Be human.'

After *Brighton Rock* came other vivid imaginings of the town. The film *Quadrophenia* was set in the 1960s and followed the fortunes of a gang leader (played by a young Sting) in the battles between Mods and Rockers. By 1980, when I went down to the seafront to watch the Mods and Rockers gathered on bank holidays by the Palace Pier, they were easily outnumbered by the crowds who had come in the hope of a replay of the great fights of the past. The famous clashes of the May and August bank holidays of 1964 had become part of the legend of emerging youth

culture, larger in the remembering. Both Mods and Rockers represented a new generation which was restless, keen to find new identities and spaces in which to challenge convention and discover new personal freedoms; they expressed their aspirations in their choice of clothing and motorcycle – drainpipes and mopeds versus leather and motorbikes. The confrontations were 'boisterous and violent but photos show laughter' and they were not vicious, claims one historian, adding that the damage to property was not extensive, although the so-called riots prompted an outpouring of self-righteous indignation. When a gang leader appeared before a Margate magistrate in 1964, the judge's tirade became famous, a final salvo in a generational battle: 'These long-haired, mentally unstable petty little hoodlums, these sawdust Caesars who can only find courage like rats, in hunting in packs, came to Brighton with the avowed intent of interfering with the life and property of its inhabitants.'

The judge – and his allies amongst newspaper leader writers – was terrified, recognising a moment of major social change; the term 'moral panic' was later coined to describe the middle-class consternation provoked by the Margate case. There was no stage more suitable on which to announce the arrival of youth culture than the seaside. Here was a place where social expectations had always been subject to experiment and transgression. The Mods and Rockers were throwing off traditional deference to their elders, eager to embrace deviancy, happy to provoke outrage; teenage culture would evolve a decade later into the punk movement, which thrived in seaside resorts, traces of which are still evident fifty years later in the prevalence of tattoo and piercing parlours, flamboyant hairstyles and fragments of edgy anarchism.

In *Quadrophenia* Sting's character offered to pay his considerable fine by cheque on the spot, as the real-life gang leader before the Margate magistrate had done. At the time, cheque books were still a privilege of those with handsome bank accounts, and the Margate incident prompted more broadsheet outrage at these wealthy rioters. Subsequent inquiries established that the seventeen-year-old involved had never signed a cheque, did not have a bank account and didn't even have the £75 required to pay his fine. A rebellious spirit and chutzpah, it turned out, could make you famous for a day and disturb a million suburban middle-class breakfast tables.

I revelled in this somewhat faded afterglow of Brighton's rebel fame, went to my first rock concerts to hear Sting in The Police, and then moved on to David Bowie, acquiring twelve of his albums and playing them on a loop on an execrable second-hand record player bought in a junk shop. In 1980 Bowie released 'Ashes to Ashes', with a music video widely acclaimed at the time, which was filmed on the Sussex coast near Hastings. Bowie takes on the seaside archetype of tragicomedy, Pierrot, the sad clown, and both the song and the video (one of the most experimental and expensive that had ever been made) were elegiac. Major Tom, his astronaut hero of an earlier hit, the lyrics explain, is now a junkie. In the haunting video Bowie as Pierrot slips below the waves, and at another point is accompanied by figures dressed in long, black, clerical dress, with a bulldozer behind: images of death, graves, burial and madness on the shore. Bowie said they were symbols of 'oncoming violence' and described the song as 'a 1980s nursery rhyme, an ode to childhood'. In a later interview he reflected that the song 'was wrapping up the 1970s, really'.

Bowie's use of Pierrot unwittingly echoed a painting by Walter Sickert made during an extended stay in Brighton in 1915: a troupe of Pierrot clowns perform on the beach in front of an audience, but many of the chairs are empty. The young men were in France fighting. The sense of an ending always hovers in the background at the seaside: the end of lives, the end of eras.

The Prince Regent set up his mistress in a handsome house in Brighton, helping to establish its reputation for amorous liaisons and infidelity. Rather than try to reform the town, Queen Victoria abandoned her uncle's palace, and built her own in the respectable privacy of the Isle of Wight. The seaside resort as a place of sexual adventure and experiment continued unchecked. Brighton developed a reputation as London's favourite venue for a 'dirty' weekend; far enough away to offer some degree of anonymity, but easily accessible. The adjective 'dirty' sums up the British embarrassment about sex, as furtive, hurried and fumbling. British literature has a rich seam of awful sex by the seaside. Pinkie's disgust is visceral and persistent throughout *Brighton Rock*, T. S. Eliot's consummation of his wedding in Eastbourne was said to have been an unmitigated disaster, and Ian McEwan's novel *On Chesil Beach* continued the tradition, telling the story of a young couple's honeymoon night in a seaside hotel, resulting in such bitter estrangement that the marriage ends and neither ever finds another partner nor resolution for their lost love. George Orwell (no stranger to secret seaside sex on his visits to the parental home in Southwold) collected saucy seaside postcards as a boy at school in Eastbourne, and in an essay in 1941 he defended their 'vulgarity' as a rebellion against convention, where 'marriage is a dirty joke or a comic disaster . . . where the newlyweds make fools of themselves on the hideous beds of seaside lodging houses'.

Brighton took on a new role after 1923, when parliament decreed that wives had the right to sue for divorce on the grounds of adultery, if they had evidence. Until the law was changed in 1969, a large number of divorces required the staging of the 'crime' of adultery, and Brighton hotels, and their maids and waiters, developed a lucrative industry to help. The 1934 satirical novel *Holy Deadlock* by A. P. Herbert described what was required for a 'Brighton quickie', suggesting that the assistance of a professional 'well-trained expert' was necessary: 'as a rule, the gentleman takes the lady to a hotel – Brighton or some such place – enters her in the book as his wife – shares a room with her, and sends the bill to his wife. The wife's agents cause inquiries to be made and eventually they find the chambermaid who brought the guilty couple their morning tea. A single night used to be sufficient, but we generally advise a good long weekend.' As the campaign for reform gathered pace in the 1960s, the Marriage Law Reform Society argued that due to 'this law of collusion, [the] atmosphere of divorce courts has become charged with subterfuge and deceit'. Absurdly, the courts became reluctant to infer sex simply from a hotel bill or a report that two people had spent the night in a bedroom together. The practice of making up evidence became well established and the Royal Commission on Marriage and Divorce took evidence from the Federation of British Detectives.

'The idea of sexual licentiousness by the seaside was largely mythical but extremely potent among holidaymakers,' points out one historian. The saucy postcards in every shop reinforced the association; with the spread of internet porn, these seaside postcards now have a charming, naive innocence, but for several decades in the twentieth century the licensing local authorities viewed them with horror, as potentially lowering the 'tone'. Some resorts banned them, others ensured they were not on open display and only sold from behind the counter in backstreet stationers. In the age of Instagram and WhatsApp, the postcards may be of less use, but the popularity of the images has only grown; they have migrated to fridge magnets, mugs,

coasters, mousepads and any other available surface. Perhaps
the most enduring of all is the image of 'Little Willie', which
sold millions throughout the twentieth century; it depicts a
pot-bellied man looking for his small son, who is standing under
his belly.

The sad story of the artist Donald McGill, responsible for
producing many of these postcards, seems barely credible
now. Four million postcards of McGill's designs were sold in
1954 alone, but the 1950s was a time of anxiety about declin-
ing national morals, and censorship boards complained in
Morecambe, Torquay, Southend, Weymouth and Brighton. In
Eastbourne the postcards were banned altogether. At the age of
seventy-nine, after decades of a quiet, respectable family life,
having discreetly churned out 12,000 designs (he ranked them
'mild', 'medium' and 'strong', the last being the bestselling),
McGill was prosecuted by Cleethorpes council in a major trial in
Lincoln in 1954, under a law which was almost a century old, the
1857 Obscene Publications Act. At the trial the elderly McGill
argued in his defence that the impropriety existed only in the
mind of the viewer, but the court disagreed, and he was fined
£50 plus costs. Thousands of postcards were destroyed, and
many businesses went bust. McGill had never kept copyright,
and profits had always gone to the printers; when he died a few
years later, he was nearly broke. A revised Obscene Publications
Act was passed shortly after, but it arrived too late for McGill.
The prints now sell for thousands of pounds.

Orwell was an early and rare public enthusiast of McGill's
postcard humour, and wrote about his childhood collection
acquired during his schooldays in Eastbourne. He maintained
that they reflected a human impulse of 'saturnalia', which he
defined as a harmless rebellion against virtue. They drew on
the humour and wordplay of the Victorian music hall, with

recurring subjects, which Orwell listed as 'sex, drunkenness, the loo, working-class snobbery, fat women, hen-pecked husbands, nervous clergymen and malapropisms'. Many followed a theme of how youth and adventure came to an end with marriage. Such humour was acceptable in the music hall, even occasionally on the radio, but, curiously, not in print, and he questioned why. Despite their vulgarity, they had an authenticity, he maintained, as 'the voice of the belly protesting against the soul', which was also evident in Shakespeare. It was a vital human characteristic which had been repressed since the beginning of the nine-teenth century, and he lamented that it had been relegated to 'ill-drawn postcards leading a barely legal existence in cheap stationers' windows'.

By the 1980s Brighton's tolerance for diversity had become a point of pride, something to celebrate and enjoy amongst its large gay and student populations. Kemp Town developed a thriving nightlife, and what was once a shabby, crime-ridden part of town is now an area of boutique hotels and smart restaurants, and is home to the biggest LGBTQ community in the country, representing 11–15 per cent of the city's population. Brighton hosts the biggest Gay Pride festival in the UK, and in the 2011 census it had the highest number of same-sex households and civil partnerships in the country. It has the only Green Party member of parliament, and a large vegetarian community. The city has a reputation for freethinking and tolerance, drawing on the many strands of its rich and varied history.

But alongside this success is another story, of a city which has witnessed deepening inequality in recent decades. Brighton has not escaped the grip of coastal poverty. Some of the most deprived neighbourhoods of east Brighton adjoin the most pros-perous. In 2011–13 (the latest available figures) the gap in life expectancy between the poorest and the richest areas was 9.4

years for men and 6.1 years for women, and there had been no improvement for ten years, the city's director of public health admitted in 2015. A fifth of children in the city are living in poverty, and many of them are concentrated in specific areas on the outskirts, such as Whitehawk, Moulsecoomb and Hollingbury. In these neighbourhoods around 30,000 people are estimated to be amongst the 10 per cent most deprived in the country, but the extent of their deprivation is often masked in city-wide data. Inequality has become embedded; one of the most graphic illustrations of this is in educational achievement. Brighton is a university city with a high proportion of graduates, but that is not true in the poorer neighbourhoods, where only 44 per cent of sixteen-year-olds achieve a Level 4/Grade C in English and Maths; in the most prosperous areas the figure is almost double, at 86 per cent. The problems set in early, with more than half of all children in areas of high deprivation starting school with a Speech, Language and Communication Need (SLCN).

The city also suffers from poor mental health, with one of the highest suicide rates in England and disturbing rates of self-harm amongst the ten-to-twenty-four age group, with a high rate of hospital admissions. It has the highest percentage of fifteen-year-olds who smoke and have tried cannabis in the country, and has some of the highest rates of teenage drinking. There is a close correlation between inequality and mental ill health, with the risk of major depression more than twice as high in less affluent areas. Shortages of affordable housing add more pressure in low-income neighbourhoods, with one of the worst income-to-house-price ratios in the country: a buyer on a low income would need twelve times their earnings to afford even the lowest-priced home. Overcrowding is twice the national rate.

This pattern of growing inequality is also evident in neighbouring Sussex resorts, and is even more challenging in

Hastings. Neither proximity to London nor the beauty of an historic resort has proved sufficient to reduce deprivation. Hastings has been the object of regeneration funds (£590 million in capital investment projects and another £90 million in social and economic initiatives), with extensive community engagement, but the town is still amongst the most deprived in the south-east. Two areas of Hastings rank in the most deprived 1 per cent in the country, and another eight are amongst the most deprived 5 per cent. A review of twenty years of the regeneration effort concluded that it had only managed to ensure that the deprivation and inequality didn't deteriorate further in Hastings. Twenty-six per cent of children in the town are living in families affected by income deprivation, compared to a regional figure

of 12.4 per cent and a national figure of 17 per cent. Even genteel Eastbourne has pockets of poverty. Nearly half of all older
people in Hastings are income-deprived, yet thirty miles up the
road in the prosperous town of Lewes the equivalent figure is
only 10 per cent. Compared to the wealth of inland Sussex, its
coast continues to be dogged by poverty and inequality.

It is 10 a.m. and is predicted to be the hottest day of the year so
far. As I get out of the car on Worthing seafront, I am hit by a
sudden surge of nostalgia for those teenage years. The Sussex
coast is instantly familiar: is it the angle of the sun, the brilliant
contrast of brightly painted terraced housing and blue sea? Or
the raking sound of the pebbles as the waves retreat? Vivid adolescent memories spring to mind; the seaside stirs the pleasures
of memory, sharp-sweet, in a way that no other place can do.

   The combination of sensations is overwhelming: the spaciousness of a wide horizon, the changing palette of colour: the
blues of sea and sky, the greens of Worthing's healthy palm trees
and close-cropped lawns, the tawny brown and soft yellow of
pebbles and sand. As I feast my eyes, my nostrils thrill to the
smell of salt, and the breeze fills my lungs; my ears are full of
the repetitive beat of the breaking waves followed by the gentle
roar as they fall back. The pebbles and sand underfoot unsteady
me, and, sitting down, I pick up handfuls of the ocean-washed
stones, feeling them like a rosary or votive beads and letting
them run through my fingers, listening to the happy rattle.
Within a short while I am in the water, and the strain of an
early-morning start and motorway traffic jams melt as I swim
through clear green water; an improvement on the biscuit-
coloured waves I remember from the 1980s. I lie back and float,
arms outstretched, and stare into a rare, pure-blue English sky.
   The development of seaside resorts in the late eighteenth

century gathered pace partly thanks to a large and unhealthy royal family. The sickly Hanoverians spilled out of London in search of cures, or at least relief, for their many ailments, and, given the significance of royal patronage for social prestige and commercial enterprise, the resorts boasted of their visits. Any royal association, however fleeting, has been cited ever since to bolster credentials. Worthing records the visit of George III's daughter Amelia in 1798. It was then a relative backwater compared to neighbouring Brighton, where the Dukes of Marlborough, York and Cumberland introduced the young Prince Regent to its new delights. Amelia's sister, Princess Charlotte, headed to Southend, and King George III went to Weymouth. In due course the young Princess Victoria enjoyed childhood holidays in Ramsgate. The Hanoverians enjoyed the seaside, although it is not clear that it did their health much good; in Brighton the dukes established a pattern of royal excess which, in time, saw their nephew the Prince Regent become obese, riddled with gout and dependent on laudanum and a wheelchair.

After a gap of several decades, the royal association with Sussex returned, as the Prince of Wales, Albert Edward (later Edward VII) drifted back, perhaps inspired by his ancestors, and at different times met at least two of his lovers, Lillie Langtry and Lady Randolph Churchill, in a house hired for the purpose in Worthing. The town appealed to this renegade heir as a quieter, more discreet option than Brighton. Later Bognor used its brief royal association to its advantage, adding the suffix 'Regis' to its name, in an early instance of resort rebranding. George V had spent time convalescing there in 1928; in response to the town's request, he allegedly replied, 'Bugger Bognor,' before dying shortly after.

Worthing has always kept a low profile, attracting elderly retirees and family holidaymakers. Oscar Wilde arrived in the summer of 1894 in search of somewhere cheap and quiet to live while he wrote *The Importance of Being Earnest*, and the playwright Harold Pinter did the same in 1962. Along with Sussex seaside resorts such as Eastbourne, Bexhill, and St Leonards, Worthing acquired a reputation as conventional and dull; Travis Elborough, the author of *Wish You Were Here*, grew up in the town in the 1980s, and is scathing: 'It was hard to come of age in a town where everyone else had gone to die ... in my late teens it was still possible to come across duffers who proudly boasted that they hadn't been to Brighton "in years" because "nowadays it was full of fuckin' queers".' By 2016 Worthing had the second oldest population in the country, just behind Blackpool, and beating Bournemouth, Southend and Birkenhead, in that order. More than one in five of the population was over sixty-five.

But Worthing had been the future once. Jane Austen based her last unfinished novel, *Sanditon*, on the town, after she visited in 1805. She portrays a seaside resort in the making, a product of property speculation, marketing and public relations. Mr Parker, who has abandoned the old family home to build himself Trafalgar House (Waterloo Terrace will be his next project), declares, 'our ancestors always built in a hole. – Here were we, pent down in this little contracted nook, without air or view, only one mile and three quarters from the noblest expanse of ocean.' Sanditon's new housing developments were 'where the modern began', comments Austen; her characters discuss how the coastline is 'full' – now one of the most familiar tropes about the English coastline. In this new, competitive, restless world, Austen's sympathies lie with quiet tradition. One character is sceptical of being rocked in a storm in Trafalgar House, perched on the clifftop, despite assurances that the storm 'simply rages and passes on'. Austen lampoons the Romantic appetite for the 'grandeur of the ocean', and satirises the enthusiasm for the bathing cure and its promise of being 'nearly infallible' for every ailment. (This last she discovered was untrue in her own case as her rare Addison's disease progressed.) But her affection for the seaside is also unmistakeable, and she was an enthusiastic bather when she visited; her protagonist looks 'over the miscellaneous foreground of unfinished buildings, waving linen, and tops of houses, to the sea, dancing and sparkling in sunshine and freshness'. She broke off mid chapter, a few months before she died, at just forty-one.

One friend at university shared my experience of living in a seaside resort. Sally grew up in Worthing, and when I ask her for memories of the town in the 1980s, her first reply was that she couldn't think of any. 'None at all?' I pressed, incredulous. She had been keen to put it behind her, and, like me, she had been in a hurry to get out, but then she remembered Mike.

'He was the lifeguard at the lido and he was gorgeous. Very handsome, clever and kind-hearted. I used to go to the beautiful lido – which has now been concreted over for redevelopment – to swim in my little bikini, and they were the happiest days of my teenage years: halcyon. Diving, swimming and sunbathing, and all the time Mike was up there on the high chair watching – he had a really good front crawl. We didn't often go on holidays as a family – my strongest memory is of a caravan stuck in the mud in Weston-super-Mare; the lido was a treat, the equivalent of a summer holiday. The days spent there felt exotic, even glamorous, my equivalent of the South of France. I lived in a shabby part of east Worthing, near to the chimneys of the Beecham factory. It was bleak, and I felt trapped in the town, but at the lido I felt free. Mike even helped get me a job as a waitress in the cafe. I fancied him so badly.

'I was desperate to get out of Worthing – Brighton was as

exciting as New York – and reach exciting places like France and London. I succeeded, and left the town at sixteen. It was a claustrophobic town to grow up in and deeply conservative. I don't think Mike managed to get away. After university he went back and got stuck; later I heard he was homeless on the seafront and it was a downwards spiral. I think there were family issues, and he became an alcoholic and struggled with depression. Tragically, he ended up hanging himself on the seafront.

'Mike's story echoed the shadow side of Worthing to me. It looks very genteel. West Worthing, with its wide avenues, broad verges and huge houses, was another world to those of us, like myself and Mike, who grew up in east Worthing. Mike is symbolic to me of that underbelly of the town, where people drank too much and didn't have decent jobs.

'Now, when I go back to visit my mother, I appreciate it. I can see it has changed hugely, benefiting from the spillover from Brighton of people looking for cheaper housing. There's even a vegan juice bar at the pier. I love the water – it's in my veins – and I don't have to get in it to enjoy it, just *seeing* the sea is inspiring and invigorating. I feel space opening up inside me. When things were difficult growing up at home and my studies were too much, I'd go to the sea and breathe deeply. It's the same now that I am going back to look after my elderly mother: it's a natural, free therapy, available to everyone.'

On my day in Worthing I experienced something similar; after my swim, I lay in the sun and listened to the sound of seagulls, waves breaking and children's cries. The simplicity of the seaside's pleasures are part of their charm. I felt the space, light and water soften me. After a fraught journey early that morning from London, navigating a family emergency and motorways, I sloughed off my everyday self for a few precious minutes that stretched into a few hours. Day trippers arrived on the beach and laid out their picnic blankets and parasols. A family with three small daughters established themselves nearby, and I watched the mother help the girls into their costumes, cover them with sun cream, while her husband lay back and sunbathed. After half an hour, she had a moment to rest. Her eighteen-month-old was fascinated by everything: the sand, the sea, and her sisters running up and down the beach to the water. She looked around her, her face full of amazement, at her parents lying on the ground, at her feet in the sand, at the bucket and spade in her hands as her sister showed her how to dig. She was being initiated into the many meanings of the seaside: how daily life was set aside, and how her parents lay as if sleeping, instead of towering over her. Every sunny day, on every beach, a child is discovering this lifetime love affair, and it was hard to drag my eyes away. Watching this toddler's astonishment was a way of recalling long-forgotten memories of my own childhood and those of my children.

The promenade was wide and smooth, and thus perfect for wheelchairs, mobility scooters, prams and bikes. Alongside it ran the busy main road as a constant companion, a steady rumble of traffic and exhaust fumes. A shelter was being used as an outdoor photography gallery – a series entitled *Beyond Land*, which 'started a month after the referendum result, with its emphasis on Britain as an island nation, geographically and psychologically separate from Europe, the photographs show a collective march to the water's edge'. The images were of a line of people following a causeway revealed at low tide out into the Channel. The label commented that it could 'also be seen as a metaphor for the times following the Brexit referendum'.

Further along the promenade, a memorial marked a Second World War tragedy, when the engines failed on a British bomber, laden with ammunition destined for Munich, as it approached Worthing. The twenty-four-year-old pilot managed to avoid the town, steering the plane on to the empty beach, where it exploded, instantly killing all seven crew. Such was the intensity of the explosion that only one body was recovered. But the audience of a packed cinema had been saved by the pilot's quick thinking and the plane only skimmed its roof; he had had the presence of mind as he faced his own death to save the lives of hundreds of other people. The young pilot had been due to get

married a week later. The 'coastal gaiety' of England's south coast is frequently juxtaposed with a brutal Second World War history; I had to blink back tears.

A little further on, I arrived at the Ferris wheel, and soon I was sailing sedately upwards into the blue sky as the figures on the beach below shrank to the size of small dolls. The view stretched out of the Sussex coast, and at the top the carriage swayed in the breeze, suspended for a few moments. England looked small and crowded, a thick coastal belt of towns and cities and their busy beaches. Only at Portslade and Shoreham did the coast become a place of industry, with the port and warehouses. To the west I glimpsed Bognor in the heat haze, and, beyond, a small patch of undeveloped coastline, where the shore reverted to being itself – an indeterminate space of shingle, sea cabbage, and a nibbled edge of sandy land. This is a stretch of coast with a particular place in the national imagination, created over centuries of literature, films, paintings and photographs. Part of its identity has always been to serve the capital with its diverse, rich offer of sex, entertainment, release and freedom, as well as fresh air and quiet for the sick and convalescing, a gentler pace of life for the retired, care homes for the elderly, and boarding schools for middle-class children.

Liam Browne landed up in Worthing by chance. Born in Derry, Northern Ireland, he moved to London and then to Brighton, and, unable to afford a place to live, moved on to Worthing. His wife is Worthing born and bred. He admits he is 'not desperately fond' of the town. 'For someone from Northern Ireland, southern England is not a comfortable place to be. It's easier in the north. I'm used to people striking up conversation, but there is a wariness here. It's just a place where people have happened to end up; I don't see the same sense of local pride which I knew in Derry.' As a child, he went to Donegal for seaside holidays; after its wild sandy beaches, the 'paraphernalia' and pebbles of the English seaside resort were a surprise, but, that aside, he acknowledges that 'down by the water's edge, the differences fall away and you are confronted by something elemental'.

Liam is self-employed, and, living on the seafront, he watches the comings and goings of the Worthing beach from his desk. He marvels at the English appetite for the sea. 'Something lures them there, even to places that are not very attractive. I think the English feel a sense of freedom at the coast that they don't experience anywhere else – it's a place where they can be themselves. I'm speculating, but it's possible that the enclosure movement [when, over a period of four hundred years, private landlords enclosed and privatised common land with hedges and walls] meant that rural England was not "theirs", and only at the seaside could they retain some sense of ownership.

'The seaside is a liminal space and it allows people to throw off constraints of time, behaviour, dress. It represents freedom. I would describe the beach as a "thin place", where the sense of something other and much larger than oneself is strong. Something fundamental changes on the beach – whatever that is. You have to be attentive, to make the effort to experience that "thinness". I try not to take it for granted. I love it when the tide is out and the light is shining on the sand; it has this extraordinary gleam. At night, the red lights of the wind turbines out at sea wink. Living beside the tide coming in and out reinforces a certain rhythm, which has a pulse; it's captured in that opening passage in *Moby Dick* – that part where everyone in Manhattan is drawn to gaze at the water.'

Later I looked up the passage Browne mentioned.

What do you see? – Posted like silent sentinels all around the town, stand thousands upon thousands of mortal men fixed in ocean reveries. Some leaning against the spiles [sic]; some seated upon the pier-heads; some looking over the bulwarks of ships from China; some high aloft in the rigging, as if striving to get a still better seaward peep. But these are all landsmen; of week days pent up in lath and plaster – tied to counters, nailed to benches, clinched to desks. How then is this? Are the green fields gone? What do they here?

But look! here come more crowds, pacing straight for the water, and seemingly bound for a dive. Strange! Nothing will content them but the extremist limit of the land; loitering under the shady lee of yonder warehouses will not suffice. No. They must get just as nigh the water as they possibly can without falling in. And there they stand – miles of them – leagues. Inlanders all, they come from lanes and alleys, streets and avenues, – north, east, south, and west. Yet here they all unite.

Worthing seafront offered Browne a troubling perspective on the pandemic: 'The beach was the one place where people could continue to go, and it was disturbing to see police patrolling and approaching groups which clearly did not belong to the same household to question them and break them up. It seemed so wrong, when the beach has historically been about freedom. The pandemic cut a huge section of the population off from the sea, with bans on travel. When lockdown was eased, the rush to the seaside was overwhelming; rubbish was just dumped – there was no sense of caring for it.'

His favourite time on the beach is at dawn on a low tide: 'A surprisingly large number of people are out then, and not just the dog walkers. It's absolutely wonderful, and what a way to start the day! You always see new things, a whole world that exists for itself, and it has a way of unsettling you for the rest of the day. In winter the English beach can be very austere, and only the locals are left, hunkering down together – it's not obviously beautiful, but I love it.'

The sea at Bognor Regis was a flat, exhausted calm, the waves barely managing to break; instead they slid under the water's silky skin on to the beach. Heat blurred the horizon, so the air felt thick and colours melded into subdued blues, white and grey. The sea was tepid, like used bathwater, and the salt dried almost instantly, leaving my skin crusty and sticky. For the first time on my visits to seaside resorts, there were plenty of people in the water; one mountainous figure of a man had taken up position, standing in the water, the folds of his tanned skin laid one on top of the other. No one seemed to be very energetic in this heat-induced lethargy, apart from the riders of the jet skis which intermittently roared past, spewing diesel fumes. A couple of young women sunbathed in their underwear; everyone was too hot to care.

I had expected something more from Bognor. Think seaside resorts and Bognor is a name which quickly comes to mind, but it is a shadow of its former self: a few seafront blocks of hotels, a small fair, a bouncy castle, a sadly truncated pier, and a lot of seagulls. A Butlin's resort dominates the eastern end of town, but, from outside the security gates, like at Skegness, one could only guess at

the cafes, pizzerias, burger bars, brasseries, water slides and palm trees which presumably lay inside the tented compound. Bognor was hollowed out; leftovers for those without a Butlin's ticket.

At the end of the pier three men were drinking cider at the only table at 11 a.m. Beside them the chain-link fence was covered with padlocks with messages of love for ever. Someone had tied a plastic bag to the railings – an enterprising visitor supplying a make-do rubbish bin? – and it was bulging with rubbish. The pier has been repeatedly damaged by storms and, without money for repairs, it has slowly shrunk; in 2008 it lost another twenty-five metres. On the beach below, fat seagulls and herring gulls were perched on the wooden groynes. The seafront kiosks offered a limited menu of chips with fish or burger. But Bognor once had a grand history; the Royal Norfolk Hotel (they added 'Royal' without Home Office approval) was where Napoleon III stayed in 1872, with his wife Eugénie, after being deposed. Queen Alexandra and her sister, Empress Maud of Russia, also visited; one can only guess what European royalty made of Bognor.

A bandstand was boarded up for repairs. Noticeboards urged residents to take physical activity to reduce blood pressure, but they were barely legible, bleached green and white by the sun. It was recognisably the same Bognor depicted in Julian Maclaren-Ross's 1947 novel, *Of Love and Hunger*, about the fortunes of a door-to-door vacuum-cleaner salesman. Head office was in Brighton, and Bognor offered thin pickings in this version of the seaside – a financially precarious world of shabby boarding houses, anxiously ageing landladies and endless rain.

More upbeat was the novel which defines Bognor's golden age, *The Fortnight in September* by R. C. Sherriff, published in 1931. It meticulously charts the Stevens' family holiday, from the packing and the thrill of anticipation, the train journey from Dulwich to Bognor, to the fortnight at the seaside, and their return. Every detail is explored with loving attention to its ordinariness: the faded boarding house, the expense of taking a beach hut, the mother's fear of the sea, the romance of the teenage daughter, the frustrations of the eldest son with his job in a London office.

> There must have been a distant strain of sailor blood in the Stevens' veins: a strain from some old merchant adventurer who knew the sea meant freedom and power – who loved it for its giant freshness, its gentleness and strength: a tough old strain that tightened the throats of his distant children and held them silent as they gripped the rails of the promenade.
>
> For Mr Stevens and his children loved the sea in all its moods: they loved it when it lay quietly at its ebb, murmuring in its sleep – and when it awoke, and came rippling over the sands: at its full on a peaceful evening, lazily slapping at the shingle. But best of all they loved it as it was today – roaring wildly round the groins, booming and sighing in the cavernous places beneath the pier, crashing against the sea wall and showering them with spray. Every one of its thousand calls had a different note – every sound was wild with freedom.

In the regimented society of the early twentieth century, when jobs were for life, convention constrained lives in multiple ways, and deference was expected in the clearly defined social order, the seaside offered a tantalising taste of liberation for the length of the holiday. Sherriff threads this theme of freedom throughout the novel: 'But over all lay a spirit of joyful, unrestrained freedom. There were no servants – no masters: no clerks – no managers – just men and women whose common profession was Holidaymaker. Round pegs resting sore places that had chafed against the sides of tight square holes – and pegs that had altered their shape, through softness or sheer will power so that they felt no aching places at their sides.' And again: 'Mr Stevens took a deep draught of the air – pressing out his lower ribs to let the ozone penetrate to the lowest part of his lungs: these were the moments that justified every pain in life – every disappointment – every humiliation.'

The sense of freedom and escape first discovered at the English seaside is now explored in resorts across the world, from Phuket to the Caribbean. Bognor incubated tastes it can no longer meet. In a 2019 *Which?* survey, poor Bognor was rated the country's joint-worst sea resort with Clacton in Essex; top-ranked was the small quiet Northumberland village of Bamburgh, famous for its medieval castle. The middle classes abandoned most of the main seaside resorts as their tastes turned towards what was perceived as natural, unspoilt, picturesque and quiet. Or edgy. Derek Jarman's garden on a shingle beach at Dungeness, Kent, within the shadow of a nuclear reactor, inspired a new appreciation amongst the cognoscenti for neglected industrialised coasts and their possibilities for solitude and regeneration. It was all a far cry from the conviviality of the crowded resort. Meanwhile, for those still drawn to the bright lights, the forms of entertainment offered at the seaside were available on multiple television channels, the music on Spotify, and the attractions bettered at theme parks with big budgets. What was pioneered at the seaside has crept inland, and on to screens, routinised as part of everyday life.

Close to the pier, part of old Bognor was being restored as new flats. On the hoardings around the development a large photo was displayed of the comedian and actor Tony Hancock on a visit to the town in the 1960s. Hancock was beaming, his hands up in the air, dressed in a bow tie, accompanied by a stylish young lady in black gloves and evening dress – a reminder of the town's elegant past. He was in the town to star in the 1963 film *The Punch and Judy Man*, set in the fictional seaside resort of Piltdown. Hancock had written the script, drawing on time he spent as an actor in Bognor in 1947. The film is an elegy, a dark comedy of a seaside resort just as its future begins to appear precarious. Hancock plays Wally Pinner, the Punch and Judy man: 'It's not the same. I used to do ten shows a day,' he complains. 'TV has killed it. It's the violence,' he adds, as Punch brutally beats the Baby puppet with a stick, to the delight of the children. The actor John Le Mesurier (better known for his part in *Dad's Army*) plays the Sandman, another of the 'beach businesses'. The plot is anchored in the insecure seaside economy of small business, akin to that of fairgrounds, struggling to fend off the interference of local officials. 'Ever forwards, never backwards' is the meaningless logo of this fictionalised version of Bognor. In a metaphor for national decline, the town's bourgeoisie are more preoccupied by petty pretension, showing obsequious deference to a visiting aristocrat, than the resort's battered fortunes.

Several other films in the early 1960s also used the seaside resort as a prism through which to explore how Britain was changing: the old social order of empire and respectability was fragmenting. *The Leather Boys*, also partly filmed in Bognor, followed the hurried romance of Reg, a biker, and his teenage bride, Dot, to their damp honeymoon, where marital disappointment swiftly set in. The conflict between their characters spoke to larger themes of social change: Dot wanted excitement and adventure and the new consumer culture, Reg was happy to accept his traditional lot and 'settle down'. Freedom was defined as individual choices about one's life. Since the seaside had symbolised freedom for several generations, there was no better stage on which to explore how it conflicted with tradition and convention.

The playwright Harold Pinter also drew on his experiences as a jobbing actor in coastal resorts and set his first play, *The Birthday Party*, in a seedy seaside boarding house. Later, he described his stay in Eastbourne: 'I have filthy insane digs, a great bulging scrag of a woman with breasts rolling at her belly, an obscene household, cats, dogs, tea-strainers, mess, scratch, dung, poison, infantility.' Pinter said that one line of the play was the most important he had ever written: 'Petey says "Stan, don't let them tell you what to do." I've lived that line all my damn life. It's the destruction of an individual, the independent voice of an individual.'

Rapid social change in the late 1950s and early 1960s was outmanoeuvring the seaside resort with bewildering speed, and in the next twenty years it accelerated. By the 1980s, the trope of the depressed seaside town prompted an inexplicably savage fury, captured in a song by Morrissey in 1988, 'Every Day is Like Sunday'. The video of a moody teenager in a town full of genteel, elderly residents was filmed in a grey, quiet Southend; the song was said to have been inspired by the resorts of mid Wales, but it could have been any resort. The morose lyrics of a wet beach and stolen clothes take a sudden violent turn, urging Armageddon on the town they had forgotten to close down. The song concludes with a wish that a nuclear bomb will come and wipe out the tedious banality of the resort.

The writer Michael Bracewell describes the song as 'an epic hymn to the melancholy of the English seaside. Both a celebration and a lament.' But advocating nuclear bombs doesn't represent a lament, rather a destructive rage. Echoing John Betjeman's famous poem which urges 'friendly bombs' to fall on Slough, Morrissey's contempt for the tedium of small-town seaside life was savage. Hancock's resolution in *The Punch and Judy Man* was to sell up and leave town, but for Morrissey, twenty years later, that was no longer enough: now, complete annihilation was required. The product of a turbulent and politically polarised decade, it was a bitter moment of national self-hatred.

6

Torquay to Weston-super-Mare

The South-West

Everything is grey: the sea, the sky, the roads, the granite of the sea walls, even the houses perched on the steep hillside. The postcard-famous palms toss limply in the cold, wet wind. On a bitterly cold February day in 2020, Torquay is a ghost town. Half the shops and cafes are closed and the other half are empty. A homeless man sits under a damp sleeping bag huddled by a fountain, begging. A few dog walkers are on the promenade, and one fisherman on the pier. It is high tide, and the sea chafes at the sea wall irritably. In the marina rows of sleek motorboats are lined up on their moorings, a glimpse of the summer glamour that has clung on in Torquay since its earliest days, when it was the favoured holiday destination of the tsars of Russia, and earned its title as the English Riviera.

On the waterfront lies the Pavilion – a wonderful confection of balconies, arches and pillars, delicately decorated with green tiles. Opened in 1912, it's a poignant reminder of an era when elegant Torquay visitors enjoyed their cocktails overlooking the sea, and it served as an inspiration to a young schoolgirl, the town's most famous resident, Agatha Christie. But it's shuttered and badged with closure notices, a sombre presence still dominating the seafront. The struggle of the English seaside resort is hard to hide, even in the most energetic and prosperously regenerated towns. Seaside infrastructure was often built at scale to accommodate the crowds of past decades and can be prohibitively expensive to restore after exposure to decades of corrosive salty spray. The sea is an uncomfortable neighbour – its spray rusts paintwork, its tides smash into robust sea walls, high winds throw up rocks and pebbles. Things get smashed, damaged and eroded every winter, producing eloquent images of decline.

The end of land and the end of lives: on the pier, padlocks and bunches of faded artificial flowers have been attached to the wrought-iron balustrade; one installation even has plaques, beads, candles, solar lights and plaster angels. What was public space has also become intensely private. One votive offering announced: 'Daughter, You are such a loving Daughter and you bring so much joy to our lives and you have a way of bringing out the best in others. You don't hear it enough but you are a dream come true, a special treasure and you are loved more than you could ever imagine.' It was surrounded by artificial flowers and solar-powered lights. Another plaque was dedicated to twenty-four-year-old Amanda: 'A Star in Life, Now at Rest, Shining Brightly in our Hearts Forever.' Who comes to visit it? And how does their grief find expression in the midst of holidaying crowds? Am I expected to politely look away, not intruding on private grief, or take an interest, curious about the tragedy of a stranger? Am I trespassing, nosy, voyeuristic or lacking in sympathy if I walk on? I'm confused by these pier memorials.

I retreat to a cafe for piping hot fish and chips and tea (at £17.00, it was more than four times the price of the same in Skegness). The hot fat is a welcome contrast to the wet, cold weather and brings much-needed warmth. The place is empty and the waitress is a particularly cheery character; since she has few customers, we fall into conversation about her children – five boys – her bio-science degree as a mature student, and the short-age of jobs. Despite the immense effort of going to university in Plymouth as a mature student, she hasn't managed to get a degree-level job, but she consoles herself with the fact that the family is happy living in Brixham, just down the road. 'It's a great place to bring up kids.' Her husband is a chef. There aren't a lot of other jobs outside hospitality, and people don't want the care work, she admits. But the schools are good, and the kids have learnt to sail. Often in the summer they come to pick her up after work, and together they go crabbing or have an ice cream – 'How many other people could say that about where they live?' she asks and smiles. She wishes I'd come on a day when there were more people about. It was too quiet. 'When it's busy in the summer, we have 400 a day in the restaurant. I like that – it gives you something to do.' That was the Torquay she wanted me to see, not this eerie, windswept place.

Seaside resorts are always under scrutiny, constantly inspected and compared, and often found wanting by exacting standards. They must always be more than ordinary, living up to expectations of multiple kinds – of beauty, fun, entertainment and, most unfairly, given the vagaries of the English climate, sun. They are places never left to be themselves but must always accommodate larger stories, including tragedy and death; or serve as a metaphor for Englishness – its decline, decay, its stoicism, or whatever else is relevant.

Eighteen months of pandemic later, I'm back in Torquay. It has just been named as one of the places – along with the Isle of Wight – likely to be hardest hit by the long-term economic impact of Covid. Economic shocks such as recessions hit peripheral places like seaside towns particularly hard, and the recovery is often long, slow and uncertain. The core economy might 'bounce back', whereas those living on the periphery can be permanently damaged. Places, like people, can suffer from long Covid. But the sun is shining, a brilliant blue sea is sparkling in the bay, and when I order a breakfast roll at the wooden kiosk in the park on the seafront, the owner hasn't a bad word to say about the pandemic.

'Lockdown was extraordinary. Everyone came to the park for their one hour of exercise, and they all wanted coffee, and we did a roaring trade.'

His usual pattern of work is seasonal, with six months of hard work, seven days a week, from late March through to late September. After that, customers are older and they only want tea, he explains, whereas the profits are in coffee. He has worked around Torbay all his life, and his 'other half' works on another beach up the coast. Before the pandemic they spent the winter in Vietnam or Cambodia – it is cheaper than living in Teignmouth. 'You can get a decent dinner with beer for £2,' he says, beaming, as he hands me a fried egg bap, for which I pay the best part of a fiver. I see his point.

When global tourism was suspended in the Covid summers of 2020 and 2021, Torquay experienced a spectacular boom, and reached the number one spot on TripAdvisor for a UK holiday. The quayside cafe owned by Kevin Shadbolt was overwhelmed.

'It's been flat out this summer. We have sixty-four covers on the quay, and during the summer it was an hour and a half wait,' he explains. 'People got fed up, but we were doing our best, and they couldn't get food anywhere else.' He has taken a break from serving to sit down to be interviewed. He also owns a hotel, and for the last two summers he has raised his prices and still had full occupancy. The only problem was a shortage of staff after 'the Poles had gone home', adding quickly, 'That's not because of Brexit.' In peak season he and his wife have eighteen breakfasts to cook every morning, but it beats his previous jobs, driving a long-distance lorry and running a freight company near Heathrow. Over ten years ago, Shadbolt sold his house in London and bought a nine-bedroom hotel, and now he has taken on the quayside cafe with his son. Torquay has been a new start for the whole family, and the dream has worked out. He proudly adds that they were winners in 2015 of an episode of *Four in a Bed*, the Channel Four TV series where hotel owners rate each other's premises.

Our conversation is cut short when the sun abruptly disappears and thick storm clouds roll in over the bay, curtaining Paignton and Brixham to the south in thick grey. When the rain arrives, it is brutal, tipping water on to the unsuspecting visitors, who are forced to scramble for shelter under any available awning, as they hurriedly dig waterproofs and umbrellas out of bags. Within minutes, the quayside tables have emptied, rain is bouncing off the abandoned plates, and the promenade is deserted. Cafe windows steam up. In the crowded amusement arcade, I find grim-faced parents and ecstatic small children gripping their cups of two-pence pieces in front of the machines where the coins teeter tantalisingly on the edge of the tray as it shunts laboriously back and forth. Each time the coins tremble. The children are cramming hot handfuls of coins into the slots, their eyes transfixed on the tray and its trinkets, which might fall at any moment. The game is decades old, and, even when a few coins fall, you would be lucky to make 10p or 20p. The machine's appeal is powerful: I am no different from the children as my hands scramble in my purse for the last two-pence piece.

My son shepherds me round the amusement arcade, anxious to ensure I don't lose all my money too quickly. We try manoeuvring the grabbers around a hideous soft toy, their odd, jerky movements stacking the odds against winning anything – if one even wanted to – and the game rapidly eats our pound coins. He swiftly slaughters me on the air hockey, my stumbling newcomer efforts easily beaten. To complete my humiliation, he takes a photograph of me in front of the scoreboard and posts it on social media. That's me done.

Up on the cliffs above the small bay of Babbacombe, next door to Torquay, the deep pink cliffs of south Devon stretching away to the north-east, Brett Powis recounts another tale of second chances. We meet in one of the four hotels he now owns in Torquay. He had been managing private hospitals, but he lost his job – 'my face didn't fit' – and he and his wife decided it was time for a change. In 2011 he took a gamble and sold the family home in Berkshire and bought a hotel in Torquay. They got it cheap; it had been in administration and water was running down the walls.

'I reckoned hospitals and hotels were much the same sort of business. I just had to change my language – guests, not patients; check-in, not admissions; check-out, not discharges,' he laughed.

Seaside resorts were in his blood after growing up, one of six children, on the coast near Bognor. The town was booming when he was a child. 'It was lovely then. At eleven or twelve, I would cycle to the bus stop, where I waited with a trolley, and when the visitors came in, I'd carry their cases two miles to the caravan park for a few pence.'

He spent all his earnings in the amusement arcades on a Saturday. 'I didn't notice Bognor was dying after I left to go to university. Since then, Bognor has had no funding, no support – the council tried to do stuff, but they've almost given up. You almost want to cry. The theatre closed, there is no direct train from London. Butlin's was damaging – it didn't bring money into the town; it had barbed-wire fences all round it. As a teenager, I would sneak into Butlin's, and we used to joke that, once in, you couldn't get out. I still go back to visit my mum and dad and sisters.'

Torquay has proved profitable, and he is proud of what he has achieved – his wife adds that she is amazed at the risky and brave decisions he has taken.

'Our hotel is the best in Torquay, and it's now four-star – we banned children and built a spa. In 2014 I bought our second hotel, primarily for the coach trade. Then I bought the Riviera, with 145 beds for a clientele of bingo-on-a-budget, hen parties and stag parties, tribute weekends, and a night club. My eldest son is the duty manager – and what the guests get up to is ridiculous, but I'm not a snob.'

Powis has had a very busy summer and is confident that he has covered the losses of Covid, but, for the first time since setting up the business, he is worried. Staffing problems are 'horrendous', and part of the reason for taking on a fourth hotel is to ensure that he can offer staff accommodation. The Covid-induced boom in the town had seen a decline in rental accommodation as landlords switched to Airbnb to increase their takings. He used to rely on Eastern Europeans, but they had gone home. He had put up wages twice, and was now offering a minimum salary of £13 an hour for housekeeping, but was still struggling to recruit. It has made him deeply regret Brexit.

'We voted Leave; we fell for the argument that we were being controlled by the EU. I don't know if it was ever true, but it was a mistake. If the clocks went back, we wouldn't vote Leave. It wasn't racism – we just wanted to be our own boss.'

He has a broader disagreement with Conservative economic policy. The levelling-up agenda was in part about the need to shift the UK away from being a low-skill, low-pay economy and needing to improve productivity, but Powis argues that this doesn't work in hotels. 'At my best hotel, productivity is low and the restaurant staff-to-customer ratio is high, but it is very successful. At my budget hotel, the ratio is low, but it's not more profitable. Productivity is not the best way to measure the success of the hospitality sector.'

He admits he is nervous. 'You just have to roll with it. I used to think I can predict, but now I just don't know.'

Torquay has been relatively successful in bucking many of the trends of long-term decline, and the south-west has maintained its appeal as a holiday destination. Its beginnings as a resort lay in the lucrative market for convalescence and invalids; like other resorts such as Bournemouth, it promised fresh air and sunshine, when doctors could offer little else to their many tuberculosis patients. Given its relatively mild winter climate, Torquay was able to prosper from a customer base of wealthy invalids wintering in the town. As modern medicine improved, Torquay had to shift tack and promoted itself as a glamorous resort on a par with the south of France; palm trees featured prominently in the interwar advertising. It was a favourite resort for those living in Birmingham and the West Midlands and, at its peak, the direct trains arrived at the rate of fifty a day.

In more recent decades Torquay and its neighbour Paignton have slowed their decline with some diversification into high tech, and the pretty area of Torbay has plenty of wet-weather attractions and enough good hotels and restaurants to attract some of the most sought-after clientele – those with deep pockets. But there is still frustration at the lack of investment in better train links. There is one direct train a day from its traditional market in Birmingham and the Midlands, and the uncomfortable, noisy rolling stock are, absurdly, converted buses; the one line serving the south-west peninsula is precariously exposed to winter storms at Dawlish. Nor has Torquay escaped the blight of coastal poverty.

Behind the October half-term bustle on the seafront on the day I visited lies a very different town, with the highest levels of deprivation in the south-west. In the wider area of Torbay (including Paignton and Brixham), a third of the population lives in neighbourhoods ranked in the 20 per cent most deprived areas nationally. A Child Poverty Action Group survey concluded that 40 per cent of children were living in poverty. With that comes deep-seated challenges: it has the highest proportion of children and young people with a Special Educational Need in the country, but with no additional funding to cover the cost of extra services. Tragically, Torbay has a high suicide rate. Austerity has hit the council's efforts to combat hardship, and by 2018 there was already concern that the council was overstretched by more than £4 million a year, and that was before Covid.

The leader of Torbay council, Steve Darling, tells me that the local authority gets £96 million less in 2021 than it did a decade ago. The effects have been devastating, and have left the council poorly equipped to deal with Covid. 'We lost 40 per cent of our staff under the austerity cuts, and our capability to make change now is significantly less,' he says. He grew up in a B & B in Torquay in the 1970s, and has seen the town scrambling ever since to slow the decline. 'We've been hollowed out, and now we're competing for money from government funds. We got £22 million from the Town Deal for Torquay and £17 million for Paignton to reshape the town centres, but that's a fraction of what Torbay used to get every year.'

The issue which most troubles Darling is affordable housing. Like many other seaside resorts, the town never had a large social-housing sector, and the stock of affordable private rented housing has been reduced by Airbnb.

'Only 8 per cent of our stock is social housing, compared to a national average of 20 per cent. We had to commission a hotel to house two dozen homeless families last winter. In the past we have had to place families as far away as Weston-super-Mare and Glastonbury, even London. Our top priority is more affordable housing, but it's like pushing water up a hill. What social housing there is in the town was transferred to a housing association years ago. The council has now established a housing company as a registered provider and we are trying to make headway, but social rented housing is not on the national government's agenda. Building in Torbay is difficult because there are few areas left for development, and they are relatively expensive, so the affordable housing is built in places like Taunton – and that's no use to us.'

Torquay may once have been famous for hosting the tsar, but by the late twentieth century it was better known as the setting for the iconic television comedy *Fawlty Towers*. The screenwriter, John Cleese, knew a lot about run-down genteel English seaside resorts, having grown up in Weston-super-Mare. He plays the part of an irascible, eccentric proprietor of a hotel on the verge of chaos; famously in one episode (which was subsequently dropped from repeats) Cleese attempts to remain polite with German guests ('Don't mention the war'). The script captures the absurdity of a post-war Englishness still luxuriating in a complacent sense of superiority and exceptionalism, in the same tradition as *Dad's Army*, another famous seaside comedy. *Fawlty Towers* has dogged Torquay's reputation since the 1980s, and not surprisingly, as a Torquay hotelier, Brett Powis dismissed my reference to the comedy series with a flicker of irritation: 'The generation that remembers that programme is dying off.'

But the suggestion that *Fawlty Towers* has disappeared is premature. It has been a spectacularly successful British export, with regular repeats all over the world. Torquay is the model for a globally understood image of Britishness; add in Agatha Christie's books, and the town has inadvertently ended up with a cultural reach way beyond its size. Christie and Cleese offer a Britishness which, beneath the politeness, combines guile, snobbish prejudice and incompetence.

The staff of *Fawlty Towers*, a delightful cocktail of characters, have been replaced by nobody: I made my hotel booking online and had an email confirmation from Grazia, giving us a door code. No reception, no restaurant, and I came and went without seeing or hearing a soul; my son overslept and met a cleaner who finally managed to prise him out of bed at midday. She kept saying 'sorry', but her English didn't go any further, and she seemed as bewildered by finding an overstaying guest as my son was astonished to be woken.

We stayed in a place which was completely anonymous; the decor – yellow folded blankets at the foot of the bed, mugs and plant pots – was similar to countless other places in Britain, Europe, America and beyond. We could have been staying anywhere, and I had the sense of being in a loop, travelling to stay in the same placeless place. The artificial flowers and flock wallpaper of *Fawlty Towers* would at least offered some sense of character; one even felt nostalgic for the sinister gentility of Christie; the chimes of the grandfather clock announcing it's time for tea, the clink of cups and saucers, lace tablecloths and the discreet phial of cyanide.

Cornwall has a dubious reputation for mystery, mysticism and magic, crystals and King Arthur, and the shopping centre where we stopped on our way to Padstow milked the theme for all it was worth – carved wooden goblins and witches peered out from amongst the trees. Padstow, I assumed, would be more hard-headed. Its reputation is built on feeding the stomach rather than the spirit, with no less than five Michelin-starred restaurants for a population of around 3,000. But by the time my visit to the town came to an end a few hours later, the place had me mystified.

Padstow meets my criteria for a resort – just. It's not one of the big nineteenth-century working-class resorts, but it offers a contrast, a late-twentieth-century version of how a seaside town makes it big as a tourist venue. Muddy fields on Padstow's outskirts have been turned over to car parks for hundreds of cars during the October half-term rush; they were probably more lucrative than the turnips or winter greens that preceded them. We follow the line of people trudging along the side of the road past modern housing estates, through a gritty underpass, before emerging in the tiny toytown that is Padstow. Each small terraced house is freshly painted in muted Farrow & Ball, the telltale sign of gentrification, and on the doors hang polished brass knockers and, in the windows, tasteful seaside artefacts – driftwood and model boats. Cranes loom overhead and a hoarding announces the building of two-bedroom luxury flats. This is contemporary Padstow's brand of Cornish myth and magic: a property bonanza. As modest former fishermen's cottages change hands for ludicrous prices, it is as if their owners are the beneficiaries of some national slot machine churning out hundreds of thousands of pounds. For those who have grown up in or around the small town, the prices are out of all proportion to their incomes. For nearly a decade, annual price rises in the town have been amongst the biggest in the country. In 2020, during the pandemic, one Padstow postcode achieved the highest yearly house-price increase of anywhere in the UK, at over 30 per cent. For locals who thought houses were somewhere to live, it's been a steep learning curve: Padstow has become an investment, a safe place to park capital, with guaranteed returns far higher than those of the stock market. Padstow has relied on holidaymakers for its livelihood for well over a century, but its cottages becoming a lucrative part of a property portfolio is something else entirely. Poor old Padstow. Cornwall has plenty of pretty harbours and a lot of good restaurants, and many of them are better connected by road and rail, but Rick Stein's amiability as a television chef (he has three restaurants in the town) has made the place famous for being famous.

Crowds of visitors mill about the few streets. Perhaps some

of them are fortunate enough to have a booking at one of the Michelin restaurants (weeks ahead they were fully booked); the rest have to queue for fish and chips or takeaway pasties. Beyond that, it isn't clear what everyone is doing, apart from browsing the expensive art galleries and seaside bric-a-brac.

We eat our Cornish pasties sitting on the concrete harbour wall and try to work it out. The wall is lined with people like us, eating their lunch; others have no option but the kerb by the roadside. The beach is a walk of a mile away. All these people have travelled for hours, on motorways and then on slow, twisting Cornish roads. Everyone has been funnelled into the same postage-stamp-sized corner of Cornwall, most of them for the privilege of eating a pasty on a harbour wall.

In season, the town's population doubles to around 5,000, but that is dwarfed by the huge volume of day visitors – 500,000 a year. Few locals live in the town centre any more, and many have no reason to visit; they don't have much use for expensive shops or restaurants, neither of which are designed for them. The restaurants are short of staff – as the billboards advertising job vacancies at the entry into town made clear – and have long had to provide accommodation for the staff they attract from Eastern Europe. Padstow has become akin to a colony, largely serviced by migrant labour, for rich Londoners and their celebrity friends staying across the estuary in Rock, in architect-designed mansions.

The mayor of Padstow, Charlie Watson Smyth, has seen much of this dramatic change over the last twenty-five years. Well into his seventies, he has lived on a farm overlooking the estuary a mile outside Padstow since he arrived as a five-year-old. His family run a farm shop. Walking away from Padstow's mayhem to meet him, I am quickly on my own, following tracks through fields planted with winter greens. Down below, the ferries to

Rock chug back and forth across the wide estuary, loaded with passengers. Watson Smyth arrives in a muddy Land Rover to pick me up, ruddy-faced, smiling, accompanied by his dog. On the way back to his magnificent old farmhouse, he swerves into a field to show me the view. The land shelves gently down to the beach, with the mouth of the estuary ahead and the wide horizon of the Atlantic beyond. 'The best view in the world,' he declares, and he points out a terrace of old cottages once built for boat pilots near the beach. 'In 1970 you could have bought one for £60k; they each have 2.5 bedrooms. One sold at the beginning of this year for £1.2 million,' he chuckles. 'The guy who sold it was born and bred Padstow, fisherman all his life. He couldn't believe it, and he told me how he had grown up poor, when most people only had an outside loo. Another cottage, which is a bit larger, with three bedrooms, sold for £3.2 million.' He shakes his head with bemusement.

Sitting in his kitchen beside a tall antique dresser laden with china, Watson Smyth is emphatic that Padstow has always benefited from tourism: 'We have grown up with the industry, and we appreciate how it makes money for the area – all the car parks, holiday lets and farm shops – the farm itself doesn't make enough money.'

His wife, Jane, agrees, born on a farm four miles away. 'My mother ran a bed and breakfast with evening meal when I was a child. We had lovely families who came back year after year, and now we have two holiday cottages here on the farm, and we see the grandchildren of the first people that came. That sort of holiday hasn't changed.'

But alongside these regulars are a new breed of visitor, adds Watson Smyth. 'What has changed is that the fortnight-a-year type annual summer holiday is a thing of the past. Now visitors come here after they've been to Barbados or skiing or Spain for two weeks.' He adds with a wry smile, 'As farmers, we always look at other people and think how much money they have. We have a lot of second- and third-home owners, and there is a lot of feeling against that, but if you speak to any plumber or electrician, they're happy. People have sold gardens and garages to be built on, and now they complain. They have made money, sold up and moved inland a few miles, where it is much cheaper. There are two sides to everything.'

There was a time when they knew everyone in Padstow, but no longer. 'In August the town is unbelievable. You can hardly move, and you certainly can't get a car through, but businesses make good money out of it. Rick has brought a huge amount of money to the town; some people say that that's pushed prices up, but others have made a killing.'

Jane has just sold a cottage in the centre of the village. It went for the asking price the day after it was listed, to a buyer with a building company in Manchester who bought it unseen.

The town council is the richest in the country thanks to an income stream – largely from car parks – of £800,000 a year; Watson Smyth has overseen thousands of pounds distributed to local community organisations. But on the bigger issues, such as affordable housing, the council has made little headway. 'I was keen on second- and third-home owners investing in housing projects, but I couldn't get anywhere, even though many were happy to contribute. Some have been coming here for years and want to help. Most of them would pay double or triple the council tax, and can afford it, but others who have retired here feel they have bought their little bit of heaven and want things to stay the same.'

One of the most contentious issues is whether new build should be reserved for local people, and when I ask his view, he hesitates. 'My Thatcherism makes me feel you can't buck the market.' He pauses, and then acknowledges that that is precisely what has happened in Guernsey and Jersey, where the property market has been split into a local and an open market, giving preferential access to locals. 'At public consultations there is always someone who comes and bangs on the table to say all the visitors should be thrown out, and then there are some visitors who are very dismissive of locals. But without the tourism, you are dead in the water. Besides, who's local? Where's the boundary? Ten or fifteen miles? People who moved here thirty years ago want to be local.'

The actor Tom Cruise was spotted in the area in the summer of 2021, and David Cameron holidayed there as prime minster and visited Watson Smyth's farm shop. The wealth passing through has become the stuff of modern-day fables; celebrity chef Gordon Ramsay bought a £4.4 million holiday home, only to pull it down and build a larger one, complete with a swimming pool with a glass wall looking out over the estuary. He can swim underwater with a view.

In a string of small resorts in the south-west, such as St Ives, Lyme Regis and Sidmouth, the rate of second (or third or fourth) homeownership can be as high as 70 per cent of the housing stock, and the property boom has spread to the formerly less well-known north Devon coast. The villages of Braunton and Croyde saw a 22.5 per cent surge in property prices in 2020. The stunning stretch from the dunes of Braunton past Baggy Point, owned by the National Trust, to the long beach at Woolacombe has become well known in recent years after the *Game of Thrones* prequel was filmed nearby. High cliffs face due west over the Atlantic. Properties on the dramatic coast have been demolished to rebuild houses which look more suited to California or San Tropez than England; one particularly lavish project won national fame on the television series *Grand Designs*.

By this point in my journey, the issue of housing had emerged as one of the key determinants of the coast's character. I took a detour from my next destination, Ilfracombe, to investigate. Emma Dee Hookway had lived in Braunton all her life, and has three sons. In June 2021 she got an eviction notice from the house she was renting, just as lockdown was easing and the summer season was starting. 'I started looking for a new place immediately, but as soon as adverts for flats for rent appeared, they were taken. Some landlords asked for a minimum income of £24,000, which ruled out anyone on benefits, and others said no pets, and we have a small dog. I just cried a lot – I felt like a second-class citizen, I felt like a failure. It was horrible.'

As a teaching assistant for children with special needs, Hookway has never earned more than £20,000, and she found that the average price for a three-bedroom house in her home town was now way beyond what she could manage. She has rented privately all her adult life, but, as the area's popularity has increased, landlords have switched to Airbnb, and flats that could once be rented for £750 per month are let out for £1,500 for a week in peak season.

Hookway's story is not unusual. Average weekly pay in the nearby Torridge district is £357, substantially below the average in England of £546, and the lowest in Devon; that makes buying a house impossible in an area where prices have been rising fast. Around 3,000 households are on the North Devon Council waiting list for social housing, so Hookway was stuck. The council's informal advice to people in her predicament was to stay on after an eviction notice, wait for the county-court ruling, and then the council has to provide temporary accommodation. But Hookway knew that could be as far away as Taunton, a ninety-minute drive, and even longer by public transport; it would mean pulling her six-year-old out of school and losing the family support in her home town.

'I'm resourceful, but here was something I couldn't solve, and there was nothing I could do. Most of the time I couldn't even get a viewing [of a flat]. One local did give me a viewing, but he had had eighty applicants in twenty-four hours, and he had narrowed it down to seven. I think he was just trying to help, but I didn't get it. I looked as far as Ilfracombe and Barnstaple, but I wanted to stay in Braunton. I suffer from depression, so I needed the support here.'

In the past Hookway has prided herself on her ability to manage, taking a second job in a bar on Friday and Saturday nights to cover the bills. Living in rented accommodation was a struggle: she was paying £875 a month in her old flat, but the Universal Credit cap on housing benefit was £650, so she had to find another £225 every month just for the rent. 'None of the kids went on their primary-school trips, but I always managed to make sure we ate well. I'm a good cook and I did bulk cooking.'

In the end she was lucky, and found a flat. It was above a working-men's club and the decor and kitchen were dated, but at least it was in Braunton. The whole experience has been life-changing. She went on Facebook to describe her struggle to find a home, and found a huge amount of support, with many in a similar predicament. She launched a petition and organised a demonstration picnic to campaign for North Devon to declare a housing emergency. National media picked up her story, and coverage in the *Daily Mirror* led to her making a BBC Radio 4 programme. She has become the poster girl for the cause and, a few weeks after we met, she visited parliament to meet Liberal Democrat and Labour MPs. Articulate and approachable, Hookway was the girl next door – she could be anyone's sister, partner or friend – and she was able to make the case in a way that local politicians had struggled to do. Local activists for

Labour and the Liberal Democrats rallied to her cause, helping to fundraise for train tickets to London. Hookway has managed to finally get attention for a housing problem which has been growing for decades.

'I've never done politics before, but now I want to contact other housing-crisis areas. I want to go national. Of course I got trolled, and I hated the headline that said I was a single mother. Our local MP denies there is a housing crisis and just told me to rent, but there are people coming down here offering a year's rent in cash up front while they look for somewhere to buy. I can't compete with that. It is not just people on low incomes who can't find somewhere to live. They can't recruit doctors in our area because of the lack of housing.'

Hookway and other local activists drew up a manifesto, which included a proposal for a licencing scheme for furnished holiday lets to regulate the number in any one area; increased council tax on second homes, with the additional revenue ring-fenced for local housing development; all new builds to be covenanted as a primary residence; and a requirement that any houses sold under the Right to Buy scheme are covenanted to remain a primary residence. On the North Devon and Torridge Home Crisis website set up by Hookway, reports are logged every few days of evictions and people looking for somewhere to live. The stories are heart-rending: often disability or the special needs of a child or parent are a factor.

Meanwhile, Hookway struggled to get a hearing from another of the local MPs, Geoffrey Cox, then attorney general. In November 2021 it emerged that he had worked remotely from the Caribbean through lockdown, and that, alongside his parliamentary work, he had continued to work as a barrister representing the British Virgin Islands government in a fraud case brought by the British government. His legal earnings had come to £955,000 in 2021 alone, and the story prompted a public outcry against MPs with lucrative second jobs. Asked to defend himself, Cox insisted that it was up to the voters of Torridge and West Devon 'whether or not they vote for someone who is a senior and distinguished professional in his field and who still practises that profession'. He has one of the biggest Conservative majorities in the country. Hookway has stepped into a political battle against entrenched Conservative complacency at constituency and national level. This is a modern-day story of David and Goliath.

The poet Ted Hughes once took his American wife Sylvia Plath on a trip to Woolacombe Bay. It was a wet November day, and the poem that he later wrote, 'The Beach', describes how Plath wouldn't leave the car but sat impassively staring at the grey sea as the rain fell on the car roof. The appeal of the English seaside can be hard to explain to foreigners. Rain has always been a part of the seaside experience, requiring considerable stoicism; shivering in a downpour or sheltering behind a windbreak. The unreliable weather drove the resorts' inventiveness, finding ways to keep people entertained while the rain poured. Mediterranean resorts have never been under such pressure to entertain – the sun and the beach are sufficient. English resorts prospered as long as the required seaside experience was predominantly about fresh air and seawater, but once sunbathing became fashionable in the 1920s, the gap between the ideal and the actual experience became all too evident. Advertising for resorts in the south-west, such as Torquay, promised sunshine, but, as I discovered on my visit, heavy rain appears with little warning.

Many a day, a bank of grey cloud sits heavily over a grey sea, as on Hughes and Plath's ill-fated visit to Woolacombe in north Devon. Other times, the coast becomes a place of spectacular drama and danger, as waves surge over railings and rear up six metres or more, to crash down on houses built close to the shoreline, scattering pebbles and sand, even snatching those who come too close, fascinated by the sea's power and strength. The British Isles, densely populated and highly urbanised, are on the edge of one of the world's great oceans, the Atlantic, with its scarcely explored depths, its myriad life forms and the immensity of its movements and currents. In Cornwall and north Devon the winter batters the coastline with huge waves, beloved of surfers, rolling on to the beaches and rocky cliffs. A visit to a resort such as Bude is a small reminder of the wildness which is our domineering neighbour, and, for those with the appetite, it provides an exhilarating wrestle with the waves. I swam in the safety and calm of the tidal pool and left the surfers, their black silhouettes like ants, racing and tumbling in the foam.

I arrive at Ilfracombe under an ominous sky, dark with rain. There is a brief moment to watch the crash of waves on the jumble of black rock on the town's small beach and to walk the dramatic headland which rears out of the town's heart, before thundering rain sweeps in. Drenched in a matter of minutes, I run to a crowded fish and chip shop and sit shivering. The windows steam up as the diners' clothing slowly dries out; a family at the next table are all on their mobile phones – they have brought their entertainment with them. The cafe is draped with spray-on Halloween spider's webs, plastic pumpkins and witches on broomsticks, but the fish and chips and tea arrive swiftly. Feeling begins to return to my cold and wet fingers and feet. Once I've eaten, I race back through wet streets to the car to peel off wet jeans and change into dry clothes.

Ilfracombe is only a few miles up the road from well-heeled Torridge and Woolacombe, but is strikingly less prosperous. At the opening of the Bristol Channel, the geography of its north-facing beach is against it; without the sandy beaches, west-facing sunsets and fine Atlantic vistas, it has struggled to hold on to its visitors. The core market fuelling Ilfracombe's growth in the late nineteenth century came from the towns of South Wales, whose miners crossed the estuary by paddle steamer with their families. By the 1960s those services had been scaled back, hit by the spread of car ownership, which allowed a wider range of holiday destinations, and by the rise of the package holiday; by 1980 the last ferry link to South Wales had closed down. Ilfracombe had already lost the railway to Barnstaple (and thus to Exeter and mainline trains to the Midlands and London) in 1970, and, without these transport connections the town didn't have a chance. What had been an hour's boat trip from Swansea now took 3.5 hours by road, or more than six hours by train (a journey which included three changes and a bus). The coastline of north Devon and west Somerset is one of the most sparsely populated areas in England, and has some of the worst transport connections.

The possibility of reviving a regular ferry service came tantalisingly close in 2010, when a new Severn Link passenger catamaran ferry was proposed between Swansea and Ilfracombe, but investors pulled out and it collapsed. Since then various other proposals have come and gone without success, despite widespread recognition that it would help turn around decades of decline. The car journey from the M5 motorway to Ilfracombe is an hour and three quarters. It's at the end of the road. As tourism declined, the hotels and boarding houses closed down and Ilfracombe found itself with a stock of relatively cheap property, a rare thing in picturesque north Devon.

What followed is familiar. People drifted to this edge when their lives became troubled, and parts of central Ilfracombe are amongst the most deprived 5 per cent in the country. The average income of £20,587 is well below the regional and national average, and there are high levels of unemployment and disability. The number of Universal Credit claimants was twice the national rate before the pandemic. The town has one of the worst affordability ratios (income to house price) in the country, and in recent years this has worsened as prices have increased faster than the national average. A six-bedroom Victorian semi-detached house with sea views and an acre of woodland for £750,000 or a two-bed luxury flat overlooking the harbour for £295,000 are tempting as second or retirement homes for Londoners, but they are well beyond the pay packets of local residents. The town has been hit hard by the pandemic yet, to the fury of the local council, it did not meet the criteria for government recovery funds announced in 2021.

The person who tried to put Ilfracombe back on the map was the artist Damien Hirst. He donated a metal statue twenty metres high: 'Verity' stands naked, her pregnant belly cut open to expose the baby within, brandishing a sword above her head, and holding in the other hand the scales of justice. She towers above the harbourside, half human, half robot, her skin peeled from her skull and the musculature of her body. She is loved and loathed. Hirst also bought several quayside buildings, opened a cafe and proposed an eco-village on the outskirts, but the love affair faded; he didn't get planning permission for his building development and in 2018 he closed the cafe. Only 'Verity' survives from his patronage. In the driving rain, I peer up at her forbidding form.

Driving the forty-odd miles to Minehead along the coast would be relatively straightforward, I assumed. It was not. The rain was still heavy and I could barely see the sinuous turns in the road; the one-in-three gradients uphill had my forehead almost touching the windscreen in my instinct to stay vertical. This was the point at which it really struck home how cut off

Ilfracombe is, perched too far north of the main roads heading to the south-west, and too far west for those venturing along the Bristol Channel coast. It takes determination to reach Ilfracombe. Defeated by the gradients at Lynmouth, I left the coast road in search of a less intimidating route and headed south across Exmoor, just as the petrol gauge swung abruptly into red. Exmoor has one of the lowest population densities in the UK, and the sheep loomed up out of the grey blur of rain, their eyes blinking in the headlights. I lost mobile connection.

By the time I eventually reached Minehead, the place seemed a welcome metropolis. My bedroom in a pub on the seafront was nothing short of a miracle: warm, dry, with grey and pink flock wall paper, a working kettle and crisp white sheets. Outside, the rain lashed the muddy beach exposed by the retreating tide. In 1999 Minehead's beach was washed away after an extensive project designed to avert exactly that possibility – the ultimate symbol of decline for a seaside resort when it is abandoned by its greatest natural asset, sand. It has since been restored for the second time in the part of the bay alongside the famous Butlin's holiday camp.

Like Ilfracombe, Minehead has suffered from the poor transport links, and the West Somerset town has the unenviable reputation of having the oldest population in the country, with 35 per cent over sixty-five, and by 2028 that figure is predicted to rise to over half. The district of West Somerset had the highest median age of 51.7 years (in 2019 the local authority was reorganised as Somerset West and Taunton). Twenty per cent of the population have no car, which severely restricts access to jobs. Fifty-eight per cent of all employment is in hospitality and food, and that means that income levels are even lower than in Ilfracombe (median earnings are £17,233, compared to a UK average of £24,006). Yet the area has another national record – it has one of the highest proportions of second homes and empty homes, owned by people attracted by the beauty of Exmoor.

Boris Johnson's family have long had a country house here, and he used to visit as a child. His sister Rachel Johnson gave an interview about those summer holidays to the writer Ysenda Maxtone Graham for her book *British Summer Time Begins: The School Summer Holidays 1930–1980*. Johnson described to Maxtone Graham a holiday regime of benign neglect in which the Johnson children along with those of other families staying in the farmhouse were largely left to their own devices. Often they were hungry, and Johnson recalled one occasion when they resorted to heating up leftover digestives in the Aga in a bid to make them edible. Maxtone Graham includes a photo of the children acknowledging that while they look happy enough, they also look rather grubby. That was how some of the upper and middle class did their summer holidays in the 1960s and 1970s, as a number of trends converged: conventional respectability lost its hold, domestic service vanished, and the overstretched mothers, who were still expected to produce large families, opted for a Bohemian collapse of standards on holiday.

Jess Prendergast's parents rented a cottage from the Johnsons for a while and sometimes came across the blond-haired Johnson children before her family moved into buildings adjoining a BBC transmitter station to develop a zoo. Her parents had spotted that the surplus heat from the transmitter was perfect for housing reptiles and amphibians, and the venture is still going strong (although no longer owned by the family). While the Johnsons went to some of the most expensive private schools in the country, Prendergast and her sister Naomi went to a comprehensive in Minehead. They got good A Levels and went on to university, but they were unusual; West Somerset has the lowest social mobility in the country, beating even the many other seaside towns which fare badly, such as Scarborough, Blackpool, Torridge (Barnstaple), Hastings and Great Yarmouth, in a pattern of poorly connected coastal and rural areas over-represented as the worst 'coldspots'. In its 2017 report, the government's Social Mobility Commission drove the point home in its foreword: 'Britain's social mobility problem is not just one of income or class background. It is increasingly one of geography. A stark social mobility postcode lottery exists today, where the chance of someone from a disadvantaged background getting on in life is closely linked to where they grow up and choose to make a life for themselves. There has been much focus in recent years on the divisions of income and class that exist in our country but far less on the geographical divide in opportunity. Our focus is on the place-based social mobility lottery.'

Social mobility was assessed across a range of measures by the commission, and on two counts West Somerset was the worst in the country. It had the lowest proportion of disadvantaged children reaching a good level of development at age five (just 30.5 per cent, compared to 69 per cent in Lewisham), and that underachievement is then set for the rest of their education; only 27 per cent achieved the expected standard at Key Stage 2. At GCSE the attainment score was thirty-five, well below the national average of fifty. If you are born into a low-income household in Minehead, the odds are stacked against you from the start. Transport links to further education are poor: the nearest options in Bridgwater or Taunton are an hour and a half bus ride away. Universities are even further away, in Bristol, making it impossible to live at home while studying. In 2018/19 the proportion of West Somerset pupils going into higher education was 23 per cent, half the national average of 47 per cent.

The Social Mobility Commission also ranked West Somerset the worst district in the country for levels of pay and affordability of housing. More than four in ten people in West Somerset earn less than the voluntary living wage (£8.75 an hour), compared with a quarter nationally. As the report noted, the three sectors that make up the majority of low-paid jobs – restaurants and hotels, retail and wholesale, health and social care – dominate seaside resort employment. In West Somerset they account for 52 per cent of jobs, compared with 36 per cent nationally, and only 24 per cent in a prosperous town such as Wokingham.

'All coastal areas were in the bottom decile for working lives, with low pay, higher rates of unemployment and weaker economic growth,' concluded the report. 'All suffered from poor transport links, critical in connecting people to jobs and wider services.' It went on to point out that an historic distortion has seen the south-east and London swallow up the vast majority of infrastructure investment: the south-east has 22 per cent of England's motorway network, six airports and three major ports; London accounts for £1,943 per head in planned government spending on transport projects, compared to £212 in the south-west, and just £190 in Yorkshire.

Disturbingly, the local government reorganisation in 2019 merging the district of West Somerset with the more prosperous and better connected Taunton Deane means that the data on the deprivation of the north Somerset coast will be masked by the wider area; evidence of this social mobility 'coldspot' is fading from view like an old sepia photograph.

The Butlin's camp dominates the shoreline of Minehead with its distinctive big-top-style tent. The other guests in the pub where I was staying – Brian and Sharon, from Sheffield – were moving on to Butlin's for a three-night punk tribute weekend for £120, to be split with two other friends. (At £20 per night per couple, it must be the cheapest holiday accommodation available in the country.) They had been in Skegness Butlin's a fortnight before. From the window of my room, the lights decorating the tents of Butlin's glowed in the rain from across the bay; I thought of Brett Powis's comment in Torquay of how the Butlin's of his youth in Bognor had been hard to get into, but even harder to get out of. I had failed to get in, despite pretty persistent badgering of the public-relations department; they had insisted that they were too busy ensuring that their customers had 'a fabulous holiday' to allow me a two-hour visit. They know their market, and it seems that did not include me, or, dear reader, you.

Nine months later I was back at the gates of Butlin's in Minehead. This time I was booked for a three-day weekend, and was ushered through the drive-through check-in. It was the Platinum Jubilee weekend, and the 170-acre site was packed with families with small children. My room, No. 3 Tributary Way, was in a long row of old-style barrack accommodation probably dating from the camp's opening in 1962. The windows overlooked a car park and the concrete walkway, but, other than the late-night revellers and the occasional marital dispute amongst my neighbours, it was quiet, and I settled in for the weekend. Everything I could possibly need was within a few minutes' walk: supermarket, pubs, restaurants, cafes, shows, discos, fairground and a huge swimming-pool complex. Even a chapel.

Intrigued by the idea of a chapel in the middle of a holiday camp, I set off to take a look. That took me on a walk encompassing most of the site. I was in the cheapest accommodation; pay extra and you get hedges and towels, perhaps even a patio, and for the highest prices (£1,633 for three nights) you could stay in a clapperboard chalet painted in muted pastel, which looked out over landscaped ponds planted with irises, reeds and water lilies. At the furthermost edge of the site were shabby blocks designated for staff; their window ledges were lined with plants and framed photos; for some, this was home. Cars were banished to a car park, ensuring that the place was safe for children. All the paths converged at the Skyline Pavilion, a vast big-top tent incorporating the amusement arcades, fast-food outlets, trampolines and other kids' play areas. The volume of pop music increased over the course of the day, but at breakfast it was still possible to have a conversation.

Sue has been coming to Butlin's for over twenty-two years, and she had come for the weekend with members of four generations of her family. 'We come for the kids,' she says without hesitation. At that point we are joined by her nine-year-old son and two of her grandchildren. 'It's always Butlin's. We've been to all three of their sites, but Minehead is our favourite. We tried Pontins once, but we had to pay for extras. We want value for money.'

Sue works in a cafe and lives near Birmingham. Her daughter, Lorraine, estimates that she's been coming to Butlin's for forty years, and adds that it's changed – it's not what it used to be. She's right. Tannoys don't wake you up in the morning any more, and the famous Butlin's Redcoats, the cheerful hosts whose job it is to boost spirits and encourage participation, are hard to find. If you have a query or need help, visitors are pointed to the Butlin's app and a telephone number for a call centre. Apart from cleaners, security men and those serving in the restaurants, there were almost no Butlin's staff in evidence. It was uncanny how the whole huge money-making machine had no personality; it seemed to run itself on autopilot. The task of chivvying people into having a good time is relegated to a clumsy slogan: 'Butlin's, the Home of Getting Stuck In'.

Sue's family have other holidays, and, as we sit over coffee, they list the destinations – South Africa, Australia, Spain; Butlin's was just for half-term weekends. Sue's twenty-two-year-old daughter, Jade, a student at Bristol University, has been coming to Butlin's all her life, and is also enthusiastic. She is planning to come on an adult-only weekend with friends. Eight-year-old Siena, Jade's niece, climbs on to her lap and intervenes in the adults' discussion about how much there is for the children to do. 'Too much,' announces Siena. Her comment is dismissed by her aunt and grandmother as her habit of complaining, but I sympathise with Siena. Butlin's has long been famous for keeping children busy, and one acquaintance reminisced to me that her childhood holidays at Butlin's ran according to an exhausting schedule. The activities themselves may have changed, but the ethos has remained. Zip wires, fairgrounds, dodgem cars, go-karts, adventure playgrounds, archery, pottery-painting, soft play, spectacular water slides in the swimming pool: the list of attractions is long and continues into the evening, with films, dance shows and circus acts. At many of these activities, parents and grandparents gather at the perimeter fences, their eyes glued on their children, encouraging them on, taking photos and filming. For the adults, all the pleasure is vicarious, watching their children enjoying themselves. But perhaps Siena is exhausted.

Sitting at a nearby table is Steve, with his wife and parents-in-law, Anne and Robert. Steve came to Butlin's as a child thirty years ago, and has been coming ever since; now his sons, eight and ten, are having much the same holiday he had as a child, and he smiles wryly as he sums it up as 'the amusement arcade and too much junk food'. All the adults agree that the weekend is about the children. 'We want to watch them grow up,' says the grandmother, Anne. 'They're not children for long.'

The Butlin's formula works for thousands of families. On my way back to my chalet one evening, I follow a weary group heading to bed, and the parents and child are all convulsed in laughter at a shared joke. They are enjoying their holiday. Over the weekend, it is evident that children are enchanted, and their dedicated parents are satisfied; it doesn't look much like a holiday for the latter, but that is not what they paid for. Before Covid, visitor numbers to the three Butlin's camps were just short of 400,000 a year, and annual turnover has risen steadily for more than a decade to reach £241 million in 2019, when £102 million was paid out in dividends by Butlin's owner, Bourne

Leisure Holdings, which owns several other big chains such as Warner Holidays. A large chunk of those dividends went to just three immensely wealthy families with substantial stakes in the holiday company. In 2020 the American private-equity giant Blackstone bought a majority stake in Bourne Leisure, and Butlin's was put up for sale less than a year later. In September 2022 it was reported that it had been bought back by a former owner, the Harris family, for £300 million.

In this merry-go-round of corporate owners, little of the profits filter down to the town of Minehead. The camp may employ several thousand (the most recent publicly available figure was 3,571, about 20 per cent of those employed in the town), but the wages are low. Cleaning and catering staff are paid about £10 an hour, and a household manager is on only £1,000 a month. While former employees pointed out online that the job suits some and pay can be boosted with overtime – the working week can reach seventy hours – with subsidised accommodation on site, others complained bitterly of poor management, long hours and drunk, rude guests.

By the evening, Butlin's pubs and cafes were busy with customers, the music was throbbing from the Skyline Pavilion, there were queues round the block for the evening's show, while out of the vast Splash World complex came the daunting roar of excited children screaming. The shows – I chose Latin American and roller skating – were exhausting just to watch, the volume of music and energetic dance steps never shifting the slightly desperate fixed smiles. While prices for a coffee or chips were higher than in Minehead's high street twenty minutes away, convenience won out, and queues built up for fish and chips, burgers and pizza. In contrast, the beach on the other side of the coastal road was empty, bar five young men of Asian heritage with broad Birmingham accents playing in the waves and posing for photos.

The sun set gracefully behind Minehead's tall cliffs, but with so much to keep visitors busy in Butlin's, few were there to see it.

Butlin's is an uncomfortable neighbour: a key employer, but contributing little more than basic wages to the town's viability. It answers to management hundreds of miles away – the headquarters of its former owners Bourne Leisure Holdings were in Hemel Hempstead, while Blackstone's head office was in New York. Butlin's sale was orchestrated by the bankers Rothschild & Co in the City of London. Minehead has known for a long time that it cannot expect much loyalty to the town from such distant investors.

Having wandered around countless chalets and long rows of housing, I finally found the chapel. Whitewashed, with a small steeple and bell, it could have been in an Italian or Spanish town rather than marooned in a Somerset holiday resort. It was much more solidly built than the surrounding accommodation. An earlier ownership with a very different ethos had ensured that some effort had gone into the place: it had tall arched windows, a brick path, and the nearby beds had been planted with palms. A handsome sign declared the chapel was open for services and private prayer, but it was locked. An incongruous relic from an earlier era, it was hard to imagine any of the recent owners financing its upkeep. I could just hear the distant thump of the pop music, and I headed back to the bright lights.

Taking a break from Butlin's, I set off for a place in Minehead which particularly intrigued me. Foxes Hotel looks from the outside like any other seafront hotel, but its staff are drawn from the college on site, attended by ninety-one students with special needs, who are training for jobs in hospitality. To mark the Jubilee, they were offering a celebratory tea. Under Union Jack bunting, I was served sandwiches, cream tea and a stand of elaborately decorated small cakes. Set up twenty-six years ago by two women, in a hotel which was owned at the time by the Baptist fellowship, the hotel-cum-college has since expanded to several other properties and is now the biggest employer in Minehead after Butlin's. Emma Cobley, the principal, has just been awarded an MBE for her work. She grew up in the area and, after a gap-year job at Foxes developed into a career, she never left, doing her teaching degree and training by distance learning while she worked at Foxes. The students study English, Maths and other academic subjects, and develop the life skills essential to holding down a job. 'The founders were wonderful, inspirational women and their model was innovative,' explains Cobley. They were able to take advantage of reasonably priced property, and the quiet, relatively small resort of Minehead offered a safe and welcoming place for the students, who come from all over the country, funded by their local authorities. Cobley puts their success down to the community support, and points out that the town benefits in several ways, from students doing work placements in local cafes, and as a valued employer in a town in need of quality jobs. A long seaside tradition of therapy and education has found a new expression.

Another experiment has taken an even more ambitious form a few miles along the west Somerset coast in the small harbour town of Watchet. The arts complex of East Quay opened in September 2021, the result of eight years of planning and organising by a group of five local women, including Jess Prendergast and her sister Naomi, with the aim of challenging the area's low social mobility and economic decline. In 2015 the local paper mill, a major long-standing employer, closed, and 180 jobs were lost overnight, knocking the town's confidence badly. I had visited Watchet before, in February in 2020, and had shivered in a chilly cafe, daunted by the scale of the dilapidation, but I had not noticed the hoardings beside the harbour which already marked the site of East Quay. On my return visit eighteen months later, the galleries, cafe, workshops, bookshop and education programme were already bustling with visitors. East Quay looks spectacular – as if a giant egg box has smashed into a lighthouse – its bold stripes echo the nautical context, and huge glass windows offer the top-floor Airbnb pods a stunning view of the Bristol Channel and the town. How had a town of 4,000 people pulled it off?

The team behind the project are known as the Onion Collective, and when I arrive at 10 a.m., one of the co-founders, Georgie Grant, is at work serving cappuccinos in the cafe. She breaks off to sit down with me to recount the story of how it began with a weekly pub drink of women with kids at the local school. They began to discuss how to help the town they loved. Eight years on, they are still hard at work; at neighbouring cafe tables, other members of the Onion Collective are in meetings, their spreadsheets in front of them. Come the lunchtime rush, they switch to the kitchen and the till.

'Over the last six months we've been in here with our hard hats, scrubbing the floors, painting, all sorts. In the first six weeks we were propping our eyelids open with exhaustion as we dealt with the inevitable teething problems,' says Grant. 'Our kids feel a bit neglected, because the work has been all-consuming. I've never been so exhausted, but I have a sense of purpose and that brings joy.

'We're the outback here, so it's self-reliance as a result of neglect. Funding stops and starts. The town has a brain drain – if kids can leave, they do, and those who stay have low self-esteem, and then it leads to mental-health problems, under-age sex and so forth. The town council was reluctant to deal with the issues; perhaps they felt they were stigmatising. Watchet has always been Conservative-voting, and it reflects their stubborn values of family and "pull yourself up by your bootstraps".

'The team reached out to dozens of local organisations, including the British Legion, the town's churches, and did twenty workshops with local people. What emerged from these discussions was that the town needed a better tourism offer, which celebrated the town's cultural heritage. We knew we needed to be self-sustaining, hence the accommodation pods.

'We have provided twenty-five jobs at East Quay, another twenty-five at our other sites, and on our projections of visitor numbers we could be indirectly supporting another 150 jobs across the town. Visitor numbers are already much higher than we had planned.'

The collective raised £7 million from government funds and foundations, and that gave the project the scale and ambition the Onion Collective wanted; they hired innovative architects and planned a gallery space which could attract internationally recognised artists. They wanted to emulate the ambition of Margate, but in a place with no easy access from major urban conurbations. It is a brave project, and speaks volumes for their dedication and passion that they managed to convince the funders. They try and balance the ambition with local engagement; in the main gallery space a major American artist has an exhibition, while in the gallery below, a community project – 'What Does Watchet Mean to You in a Bottle?' – features dozens of delightful quirky contributions celebrating the town. A family see my notebook and come over to tell me how amazing the place is. 'We have been here every day for the half-term workshops for the kids – mud batik, shore safari, you name it. We're staying in a cottage on Exmoor, we come from near Bristol.'

Providing workshops for middle-class families in self-catering accommodation is not the real aim of East Quay, but at least it is bringing new visitors into the town. Getting the balance right is challenging. 'Gentrification is a really difficult flip side of what we are doing, and it could be an impact of the centre. It's the dark side of providing opportunity,' admits Grant.

The cafe had filled up with a large party celebrating a wedding between two retired residents of the town. Grant breaks off to greet locals. This is the bit she loves, she says, seeing people having a good time. 'Our aim was to create a space which offered different opportunities – creativity, culture and connecting people to the imagination and joyfulness. We started with freight containers as spaces for artists, which were refurbished by volunteers, and then created a courtyard with events and meals. Then we got funding to run kids' events. It grew and grew, and now the cafe has been at capacity since we opened. The theme of the bookshop is change and hope.'

Later, I catch up with another member of the collective, Jess Prendergast. 'You have to demonstrate aspiration, you have to be able to say, *Look!* You have to live by example as role models, you have to be relevant, so that young people think that you come from here, went to their school and did this,' she says.

Six months later, I'm back in the town to stay in one of the new accommodation pods at East Quay. I have a metal spiral staircase in the two-storey loft apartment, and I can sit in a deep armchair with a view out of the floor-to-ceiling windows over the old harbour and the blustery skies above the Bristol Channel. Downstairs there is a breakfast of mashed avocado on toast and good coffee. I'm impressed. But that's not the case for all of Watchet's inhabitants. Later, as I walk along the harbour wall, I chat with Gladys, who is waiting to watch a relative going out in his boat. Well into her eighties, she has lived in the town all her life, as do many of her family, and she has nothing good to say of the East Quay project. She would have preferred the money to be spent on new public toilets, and as for the art – well, children in the town could do better. 'It's not Watchet,' she concluded firmly, as if that settled the question. The founders had created jobs for themselves, she noted, and her complaints continued at length about the building, its contents and the cafe's menu.

It was a reminder of the enormous difficulty such cultural regeneration projects have in winning community support and engagement, despite all the hard work and dedication of the founders. The new curator, whose previous job was in Vancouver, Canada, aims to put Watchet on the international map for contemporary art; he is delighted that the gallery has already appeared in the *Guardian* three times. That's a world away from the traditions which have sustained Watchet's independent-minded resourcefulness; but Gladys speaks for an older generation, and the East Quay project is about creating a future for her grandchildren and great-grandchildren in the town. Only time will tell.

I'm back on the Bristol Channel, on a day of hot and bright sunshine. A cloudless sky arches over Weston-super-Mare's nearly empty beach. I arrived early, hoping for a swim before the tide went out, but the sea had vanished. High tide was at eight o'clock and I was too late; there was nothing deeper than the toddlers' paddling pool built close to the promenade. The rich brown sand spread as far as the eye could see – it is a mile from the seafront to the sea at low tide, and, in case I was tempted to take a hot, shadeless trek across the sand in search of a wave, the beach was dotted with signposts: a black silhouette against yellow of a flailing person, arms outstretched in desperation. Danger, sinking mud. So, like everyone else, I left the acres of beach well alone and had to assume that the distant line of darker brown was the sea: it had the appearance of chocolate sauce.

Weston-super-Mare has little to do with the sea for at least twelve hours out of every twenty-four. By the end of my visit, I was disorientated: how had the expanse of mud glistening in the white sunshine constituted the seaside? I'd heard no waves break and I'd not swum; it felt like a trip to an English desert. Perhaps that's why Weston has to incorporate the sea into its name as a reminder.

But Weston has some magnificent compensations, and, historically, they were enough for the holidaymakers from Bristol, Bath, South Wales and the land-locked West Midlands who arrived on the railways. Chief among them, to my mind, must be the lumpy outline of the island of Steep Holm, heaped on the horizon like a handsome Christmas pudding. From every vantage point, it is an unmissable landmark, an inelegant lump (privately owned) in the Bristol Channel. One cannot but love it as it draws the eye out to the horizon, away from the steady stream of traffic lumbering along the seafront. In the far distance on the South Wales coast sits a line of chimneys, docks and tower blocks, which is similar to Southend's view over north Kent. Estuary resorts juxtapose pleasure with industry; forget picturesque, luxurious or exotic, the impact is bracing and expansive.

Even the truncated pier had no chance of reaching the sea. It smelt of chips, and its rails offered a comfortable perch for the predatory gulls waiting for the leftovers. The tinny music of Rod Stewart, Elton John and other fifty-year-old hits substituted for the sound of breaking waves. Anyone still hungry after the mountains of chips could tuck into tubs of ice cream piled high with whipped cream, jam, chocolate chips, sweets and lashings of chocolate sauce. The seaside combination of indulgence and ill health – fat, sugar, obesity and mobility scooters – was by now a familiar and uncomfortable experience.

A brisk breeze whipped the tidal pool into choppy wavelets, but no one was tempted to swim in this brown jacuzzi, even an enthusiast like myself. The history of pollution on this north-eastern Somerset shore – in particular, high levels of cadmium – acted as a further deterrent. More popular were the bars and cafes, and in the sunshine they were busy with cheerful drinkers downing pints of beer.

One customer was a surprise: in the window of the Midland Hotel sat Queen Elizabeth II. The life-size model was in full evening dress and tiara, with a corgi on her lap. She looked a little bored, as if she had been there for a while, even a couple of decades. It was as if Weston had decided not to be outdone

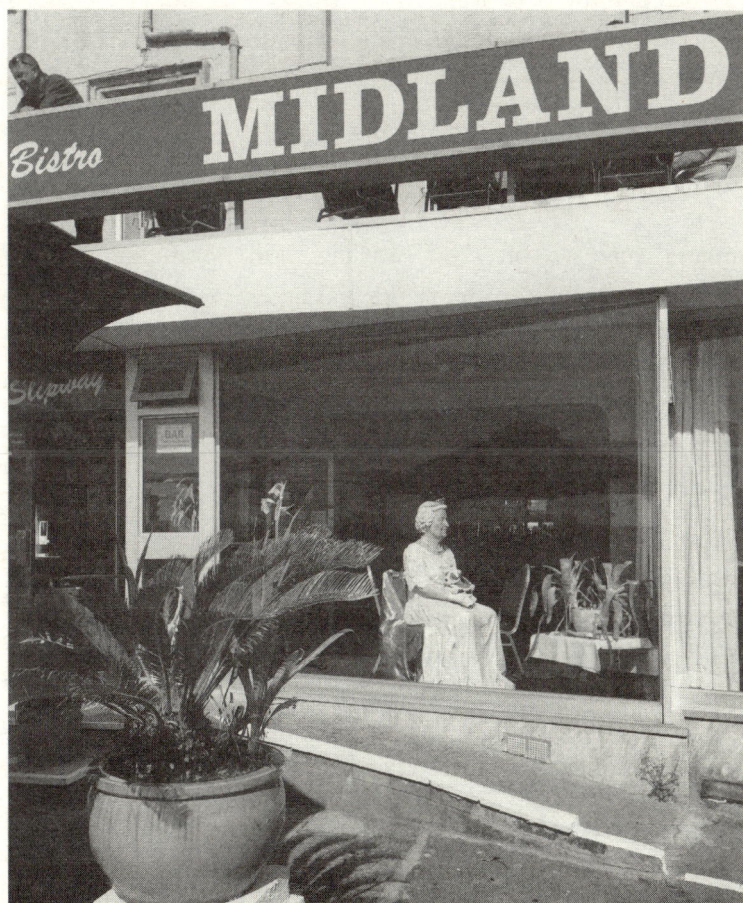

in the long history of royal patronage of the English seaside –
Torquay's connections with the tsars, and Brighton's association
with the Prince Regent – and had installed the queen herself.
No one took much notice of her, and a noisy group of visitors
on the balcony above were already inebriated, while the former
queen sat in the window in the sunshine, with her hands in
white gloves clasped beneath the dog in her lap.

At the north end of town, the grand terraces were visibly
crumbling, losing letters, guests and windowpanes in roughly that
order. Giant pineapples of carved stone perched on top of the colonnaded entrances, referencing the Bristol Channel's long, painful
history with the Caribbean. The graffiti bloomed with the fantas-
tical, mystical and the vaguely sinister: 'What you seek is seeking
you.' Birnbeck Pier was in the final stages of decay, limping across
sixty metres of mud to the old pleasure palace on Birnbeck Island.
It was wreathed in barbed wire, with signs warning of multiple
dangers, such as soft sand, sinking mud, submerged objects,
rip currents and fast-moving tides. The message to prospective
adventurers was clear: this pile of rusting iron, collapsing arcades
and rotting timber was perilous. Looking through binoculars,
I could see the island's once-elegant arcades, handsome stone
lifeboat shed and spectacular ironwork, which the Atlantic gales
were slowly dismantling. It had been a grand Edwardian theme
park, and the ruins were tantalising, brimming with ghosts and
evocative in the way only abandoned, inaccessible ruins can be.

It was enough of a struggle for Weston to maintain its Grand
Pier (a blaze in 2008 destroyed the end pavilion and the rebuild-
ing cost £51 million), and it had had to give up on Birnbeck. But
as Brighton and many other seaside towns well know, disowning
a pier is not easy; it sits there, complaining, pleading for help,
prompting campaigners to lobby as each limb drops off. A small
sign suggested that someone had taken on the gargantuan task

of trying to launch a regeneration project for Birnbeck pier – one feared it would prove a monument to a lost cause.

Weston's former boarding houses and hotels have been divided up into cheap bedsits, and, given its proximity to Bristol and Bath, it faces similar challenges to many other coastal towns, with a particular reputation for substance abuse and its treatment. Deprivation is amongst the worst 1 per cent in the country, and there is a *ten-year* gap in life expectancy between Weston's Central ward and neighbouring Clevedon, a few miles north, (sixty-seven versus seventy-seven years). The relatively cheap housing led to the highest number of drug and alcohol rehabilitation centres in the country. The number has been reduced in recent years following complaints from residents as the town struggles to curb drug dealing and county lines; convictions for drug dealing and police raids dominate the local media.

Weston was catapulted into a brief and spectacular moment of global fame in August 2015. For months hoardings had been erected around the derelict site of a former art deco swimming pool on the seafront. Locals and even workmen on the site were told it was for a film set, and only three people really knew what was happening, one of them being the chief executive of Weston council. When it was finally unveiled, it was revealed as an extraordinary pop-up art exhibition curated by Banksy, with the participation of dozens of major artists. 'Dismaland, A Bemusement Park' was a dystopia with a savage, caustic sense of humour: a lady sat on a bench, her upper body obscured by wildly flapping, aggressive seagulls, one arm outstretched in panic; in another piece, several horses were missing from a traditional fairground carousel, and a butcher was sitting on boxes labelled 'Lasagne'. Inside a derelict castle was Cinderella's overturned carriage, smashed, the white horses lying dead, and the body of Cinderella herself mangled as the paparazzi cameras flashed.

Another model depicted boats full of refugees bobbing in the sea at the foot of the White Cliffs of Dover. A shop offered pocket-money loans at 5,000 per cent interest. One display gave detailed instructions for how to break into bus stop advertising display cases with £5 packs of tools for the job. Meanwhile, visitors could warm themselves at daily bonfires of the books written by Weston's most famous son, Jeffrey Archer, a bestselling author and disgraced Conservative politician (once known as Baron Archer of Weston-super-Mare), who served a two-year sentence after a

conviction for perjury and perverting the course of justice in 2000. Visitors were greeted with airport-style security of X-ray machines and scanners crudely made out of cardboard.

Over thirty-six days, 150,000 visited (including Brad Pitt), with cheap or free tickets for thousands of Weston residents, but critics complained it was depressing. 'The gloom of the British seaside at its most dilapidated and moribund wells up in me,' wrote Jonathan Jones in the *Guardian*, adding that Dismaland was 'just a media phenomenon, something that looks much better in photos than it feels to be here. "Being here" is itself just a way of touching the magic of Banksy's celebrity – that's why everyone is taking pictures. This is somewhere to come to say you went.'

Looking back, Dismaland's savage satire is still sharp and clever. It was a witty new iteration of how the seaside has often been the site for humour – be that saucy postcards or the gentle ridiculousness of *Fawlty Towers* and *Dad's Army*. It generated a welcome £20 million boost to the local economy, but Jones's conclusion had much truth: there was no evidence of a longer-term benefit to the town from its month on the global cultural map. Banksy did not make much (if any) money out of the project, but there was a sense that, perhaps inadvertently, he had used the place, exploiting Weston's own already vivid version of dismal land – crumbling buildings, mud and drug addiction – as a backdrop. Dismaland offered no hope for a better future, only vandalised bus shelters. Millions around the world, no doubt, smiled wryly into their laptops at the wit, but it left a sour taste. It was too close to the bone.

7

Blackpool and Morecambe

Lancashire

My hotel in the centre of Blackpool is in a narrow, terraced street, lined with similar, modest, family-run establishments. In front of the bay windows with their bouquets of artificial flowers sit the guests, smoking and sometimes drinking, enjoying the quiet. It is a sunny day, so why aren't they on the beach or sitting in the sun on the promenade instead of in these shabby, dusty backstreets? What do they know that I don't?

Blackpool seafront has been grabbing the attention and cash of visitors for over a century, and its cacophony of noise and glitter is overwhelming. Light, noise and smell have been recruited to swamp the senses: brash blinking neon, air saturated with sugar and frying oil, the thud of competing music systems, and the constant rumble of traffic. The bars are packed with early-evening business. It is the first weekend of the famous Illuminations, when seven miles of seafront are strung with vivid, colourful lights, and as the dusk gathers the traffic slows to a bumper-to-bumper crawl. The cars' back windows are crowded with the eager faces of excited children, and more music thuds from open windows.

Dazzled by this largeness of life, I keep walking under the jewelled canopy of lights. One installation in flashing emerald green and purple neon shows Jesus on the cross, with the title, 'I am the resurrection and the life.' A couple are admiring its lurid beauty in the dark, and Laura tells me how she is Blackpool born and bred and has raised her children here, but how it has changed in the last fifteen years. 'There is too much drinking

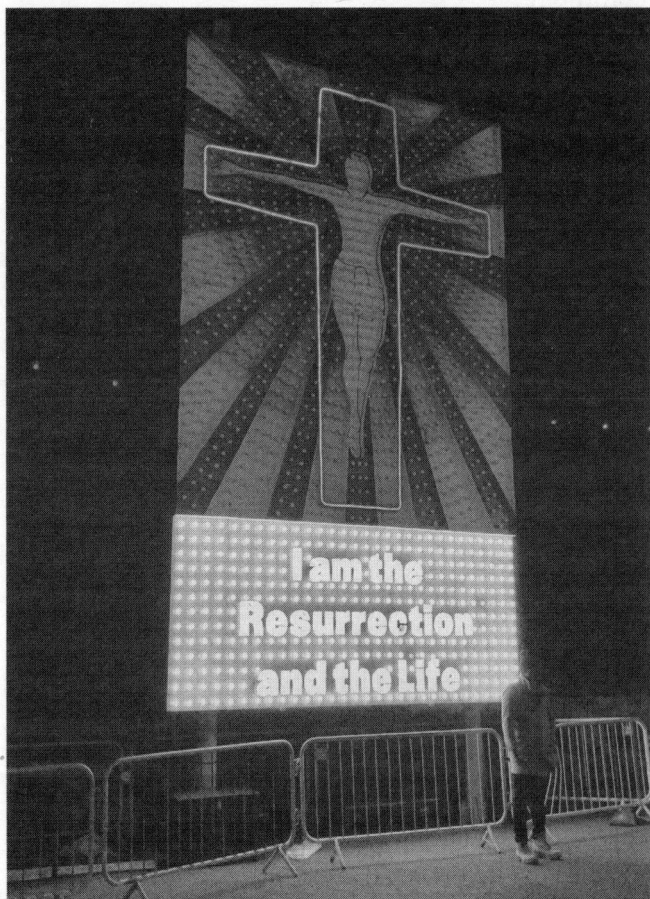

now. Everything is about the drinking. When I had my kids, the adults didn't sit about drinking,' she laments. 'And it's dirty.' She has moved to the northern outskirts, while her new partner, Moses, has retreated to neighbouring Lytham St Annes, after eighteen years in the town. He would love to live somewhere with a more diverse population, but he's got a job and he's got used to the cold – he originally came from Gambia – and now he's got 'naughty Laura', as he jokingly teases his partner of three months. They seem very much in love. Living eight miles apart, they have all the noise and drama of Blackpool between them.

In the centre of the town, the majestic Victorian Blackpool Tower looming overhead, I come across a group on the pier whose lavish outfits match the setting. Two are dressed in short pale-pink dresses, tight bodices edged with elaborate frills, the skirts showing flounces of white petticoat. They wear bonnets of pink and white, and the outfits are completed with pink patent knee-high boots. A third is in eighteenth-century black tails, white cravat and top hat. They roll their eyes with exasperation when I ask for a photo, and explain the Japanese Lolita subculture which inspired their outfits. 'Look it up,' says one young woman, who says she is from Bristol. Her companion, a skinny young man with bright-red lipstick and yellow teeth, is from Blackpool and works in a call centre. He has a dainty pink cap set at a jaunty angle on his long blond hair. He only dresses up like this 'once in a blue moon', he says, but the Bristol lady was a regular. Online I discovered later I could have matched their outfits for £60 and joined one of their meet-ups. It says much for Blackpool that no one else seemed much interested. It's a town built to cater for the outlandish and the excessive.

Samra Mayanja, the artist I interviewed about growing up in Harwich and who has been making *Edging Home*, a series of podcasts about English seaside resorts, had come to Blackpool to meet me and do research. Over supper, we caught up, swapping notes on our shared fascination with coastlines and England's edges. Nowhere intrigued us more than Blackpool, England's quintessential seaside resort and still its most popular, drawing in eighteen million visitors a year pre-Covid, numbers which bounced back quickly after lockdown. Surveying the seafront dining, the options were limited. It was fish and chips, again. At the table next to us, three young women got up to leave, their diamante sandals glittering, their slim bodies sheathed in tight dresses, their long glossy hair cascading down their backs; glamour has always been part of Blackpool.

Our seafront restaurant was only a few feet from the hooting of car horns and thumping music, but, regardless, we discussed Mayanja's recent research into a delegation from the Ugandan kingdom of Buganda, who came to the coronation of Edward VII in 1902. A missionary society funded their tour of England and published an account of their travels, translated into English. Her interest lay in how the Ugandans struggled to find words in their language, Luganda, for the things they were seeing: railways, factories, cars, iron bridges and electric light, and perhaps even such marvels of Victorian engineering as Blackpool Tower. It felt like an odd conversation to be having in the midst of this hectic holidaymaking, but later I decided that it had been curiously apt. I was struggling to get my bearings.

Blackpool has been shorthand for everything about working-class culture that the middle class judge as crude, cheap, tacky and vulgar. As a young reporter, I attended party conferences in the early 1990s in Blackpool, and the London media types bitterly complained about their shabby hotels, revolting fried breakfasts and limited options for dinner. I have never met anyone who visited the place by choice. While ties between class identity and forms of entertainment such as sport, music, fashion and television have been eroded (often by the middle class adopting working-class appetites), the seaside is a striking exception, and the class identity of resorts is starkly evident. For decades Blackpool has been trying, with limited success, to attract middle-class visitors with more money to strengthen its precarious economy based on cheap holidays; yet even professionals working in the city admitted, slightly embarrassed, that they rarely visited the seafront, despite living nearby.

When I later recount my experience of that first evening and my sense of dislocation to Vanessa Toulmin, a history professor and advisor on cultural regeneration in seaside towns, she laughs, immediately understanding the issue. Born in Morecambe, she says she experienced something similar when visiting the middle-class resort of Southwold in Suffolk. 'I was discombobulated. I didn't recognise this as my type of seaside. What do you *do* here? I asked my hosts, and they didn't even understand the question. "*Do?* Well, we walk or look at the sea."' She found their reply utterly ridiculous, she tells me, breaking into laughter again. For her, the seaside was about an exhilarating frenzy of entertainment, excitement and thrill; she wanted the buzz of crowds and music.

Blackpool invented the mass working-class resort; early visitors included the 'padjamers' who came to swim on Lancashire's Fylde coast in the 1820s – without gender segregation, usually swimming naked, having drunk large amounts of alcohol – but it was the building of the railways that brought the huge crowds. They came from the Lancashire mill towns, and the seaside trips were a reinvention of the old tradition of 'wakes weeks', when the textile mills closed for repairs and maintenance in the summer, and the population went to the local fairgrounds. The trains could take them to the sea instead, and each major mill town had its allotted week. At the time, Lancashire was one of the richest counties in England, and its coastline was in easy reach of two of the country's most important cities, Manchester and Liverpool. With this big customer base, the three main resorts of Blackpool, Southport and Morecambe grew rapidly, with huge investments in seaside infrastructure such as piers, parks, theatres, promenades and ornate winter gardens, which housed a range of attractions such as zoos and aquariums, concert halls and ballrooms. Most spectacular of all was the Blackpool Tower, built to rival the Eiffel Tower in Paris.

When the Blackpool Tower opened in 1894, it was the tallest building in the UK. Visitors were whisked up by lift to a view that stretched from the Isle of Man to Blackburn and Liverpool on a clear day. No expense was spared in the foyers and landings, with their panelling, armchairs, chandeliers and ceramic tiling of birds in blue, with polished red-brick balustrades and stained glass. The complex boasted a menagerie, which included monkeys and lions; a circus, which in 1906 included forty performing polar bears; and an aquarium. Carved beams and brightly painted pagodas featured in an Oriental arcade. The ballroom had a capacity of 6,000 guests and extravagant rococo decor, a fantasy of the Neapolitan riviera. The famous dance floor was of sprung mahogany, oak and walnut parquet, polished like a mirror. It was 'modelled on the great European ballrooms of the aristocracy, its decorative brilliance, with its illusion of privilege and grandeur, were in reality a democratic space built for working people, where a factory girl became a duchess,' writes Toulmin. In Arthur Laycock's 1900 novel, *Warren of Manchester*, the 'mill-girl glided cheek by jowl with the factory master's daughter, the spinner with the smart city clerk and the stylish salesman'. Poignantly, by 1934 the thickness of the wooden floor had been worn down 16 mm by the feet of an estimated 100 million visitors. Blackpool's long-serving former MP, Gordon Marsden, describes the Tower as a 'yellow brick road – a dreamscape where the comic, the ecstatic, the down to earth and sometimes the wistful and melancholic play out in the memory'.

What made all this development possible was Lancashire's textile wealth. Its working-class household income was the highest in the country by the mid-nineteenth century, boosted by female, teenage and child labour (only in the 1890s was the labour of children under ten banned). Saving for an annual holiday in a boarding house was within the reach of thousands of families, even if that often entailed one room and an outside privy. To help those nervous of the adventure, organisations such as holiday clubs, churches, the temperance movement, the Salvation Army and friendly societies arranged everything; holidays were with your neighbours. A generation before anywhere else in the UK, the Lancashire working class invented the family holiday, and the rest of the country in due course copied, after the first bank holiday became law in 1871.

The prime minister, Benjamin Disraeli, called it a revolution, and he was right; fifty years before, foreign commentators had been astonished by the violence and crudeness of English male-dominated entertainment, with its drunkenness, gambling, racecourses, fairgrounds, cockfighting and an 'undercurrent of disorder and violence'. The growth of the seaside resort became part of a bigger 'civilising mission' on the part of the middle class – parks, libraries and promenades were planned to reform and tame what were seen as baser instincts and inculcate better manners; fresh air might lure customers away from the pubs. Fears that the working classes would trample flower beds and break windows were gradually allayed as 'entertainment was educationalised', as Toulmin puts it. The middle-class reformers were keen to keep people busy, and hoped that in these new public spaces in cities and at the seaside the working classes would learn better manners by mingling with their 'social superiors'.

The project was judged largely successful, but it also entailed, in seaside resorts such as Blackpool, constant battles over what was described euphemistically as 'tone', when the middle-class aspirations of local authorities clashed with the habits of some visitors. Local by-laws licensing entertainments were a key battleground in which the two came into conflict. One strategy to exclude the undesirables was to impose ticketing, effectively creating different zones in the resort. Blackpool's North Pier and its surrounding area of smart hotels catered for the middle classes, the south shore for the less prosperous customers. In the main attractions, such as the Winter Gardens, where the classes mixed, vigorous attempts to uphold middle-class standards were made, although not always successfully; prohibitions on spitting, for example, had been abandoned by the 1890s. The Victorian and Edwardian seaside resort was a 'crucible of conflict between classes and lifestyles, as wealthy and status-conscious visitors and residents competed with plebeian locals and roistering excursionists', comments historian John Walton. 'The seaside brought mutually incompatible modes of recreation and enjoyment into close proximity in ways that seldom happened inland and gave an added edge to the perennial Victorian debate about the proper relationship between leisure, class, religion and morality.'

Increasingly, Blackpool served a valuable social purpose as a release from the discipline of the mills and the tyranny of Victorian respectability. Its array of entertainments distracted the mill workers from their harsh economic exploitation and

impotence to bring about political change. By 1906 three million had visited the Pleasure Beach for attractions such as the
country's first helter-skelter, a trip through river caves, the water
chute, and themed arcades such as Spanish Street. The first
roller coaster was unveiled in 1907. In the more disreputable
seafront stalls, Blackpool indulged appetites for the outrageous,
the scandalous and freakish, drawing on fairground and circus
traditions to horrify and fascinate. In the 1930s researchers were
alarmed by the gullibility of the crowds, fearing they could be
easily manipulated. Alongside the fortune tellers and clairvoy-
ants were more outlandish entertainments: people alleged to be
enduring extremes of starvation and confinement, exhibitions
of murderers and figures ravaged by venereal disease, even a
so-called 'intersex' woman purporting to be a man who had
married a woman, to the horror of contemporary commentators.
This was Blackpool at its most anarchic, carnivalesque and flamboyant, recovering those repressed traditions of working-class
entertainment which had been 'educationalised'. Even strict
rules on sexual propriety could sometimes be set aside at the
seaside; photos show young couples lying entwined on the beach
kissing, albeit fully dressed. The 'prim and the Rabelaisian
sides of British character came into maritime confrontation' in
Blackpool, adds Walton. The crowds' vibrant energy incubated
a playful adventurousness with aspects of identity, be that of
class, gender or sexual behaviour.

This new sense of freedom was captured in a groundbreaking play, and later film, *Hindle Wakes* (it had no less than four
screen versions: 1918, 1927, 1931 and 1952). Arguably the best
film ever made about an English seaside resort, and probably
the most influential, it is a magnificent celebration of everything
Blackpool represented – the exhilaration, playfulness, the
questioning of convention. The cinematography of the famed

attractions at the Tower, and Pleasure Beach's Big Dipper, and Helter-Skelter, represents the dizzying experiences of perspective, speed and gravity. In the film, Blackpool is 'portrayed as a land of dream-like happiness and possibilities', comments the critic Lara Feigel. Influenced by Henrik Ibsen's *A Doll's House*, the 1911 play portrayed a woman claiming sexual freedom in a manner well in advance of its time; the protagonist, mill girl Jenny, meets Alan, the son of a mill owner, in Blackpool, and they enjoy a weekend together. Despite parental pressure, she rejects marriage; she is as entitled to as much fun as a man, comments Toulmin, adding, 'It had a massive impact on me when I first saw it. Jenny is a Lancashire lass.'

By the late 1930s seven million were visiting Blackpool every year, out of a national UK population of forty-six million, and the town had 500,000 visitor beds; it comfortably beat the 5.5 million visitors of its closest rival, Southend. By the twentieth century 'entertainment was industrialised for the masses,' said Toulmin: the attraction was no longer primarily the sea or even the beach, but the dancing, cinema, and Pleasure Beach amusement park, with its American-style rides, each invention more dramatic than the last. One contemporary writer commented of the Pleasure Beach that it was 'the final solution of the periodical need for an orgy, a safety-valve for the high spirits of mankind'. It was 'the birth of an art of excess and garish overabundance' and 'ecstatic freedom' from the routine of urban industrial life, suggests Feigel. Middle-class commentators were nervous of this mass culture by the 1930s, haunted by Europe's move to authoritarian totalitarianism. The Mass Observation project, first set up in 1937 to record everyday life, sent researchers to Blackpool, and they were disparaging, concluding that mill workers in the resort went 'to the places where the crowds are, where the rhythm is as fast and the noise almost as great as that of the mill'. In 1934 the writer J. B. Priestley wrote about Blackpool with barely concealed distaste in his *English Journey*, describing it as 'a great roaring spangled beast' and 'the huge mad place with its miles and miles of promenades, its switch backs and helter skelters, its array of wine bars and oyster saloons and cheap restaurants and tea houses and shops piled high and glittering with trash'.

Brash and unapologetic, Blackpool's resort economy was relatively protected during the Second World War (unlike the east-coast resorts, many of which closed due to the threat of bombing) and benefited from troops stationed in the area, and it swiftly resumed business in the late 1940s. By then it had 20,000 theatre seats to fill a day. Only London was bigger, and, throughout the twentieth century, Blackpool attracted the biggest music stars while incubating new talent. 'You could start as a stand-up comedian on the pier and end up in venues with an audience of 3,000. Morecambe and Wise, Les Dawson,

Lenny Henry all got their breaks in the north-west,' points out Toulmin. Blackpool's cultural prestige was such that it was the only place outside London where Frank Sinatra played on his UK tour in 1950.

Alfred Gregory's photographs of Blackpool in the 1960s show beaches so densely packed there was barely room to move between rows of deckchairs. He caught on camera the tipping point in popular culture between the traditional amusements – donkey rides, ice cream, Punch and Judy sideshows, piers – and the emergence of a new popular culture of rock and roll, TV variety shows, and a growing permissiveness, with strip clubs and tattoo artists. The numbers visiting in the 1950s and 1960s were huge, remembers Gordon Marsden (MP for Blackpool South for twenty-two years), who used to travel there with his family as a small boy from Manchester: 'It was so busy that trains were leaving for Blackpool every five to ten minutes.' Even in the 1980s Blackpool was still the biggest seaside resort in Europe, and had more visitors than the whole of Greece and more hotel beds than Portugal. As Keith Waterhouse put it in 1989, Blackpool was 'out-vulgarised only by Las Vegas'. Even today, its visitor numbers, more than any other seaside resort, put it amongst the biggest attractions in the country. Marsden points out that 'whenever there is a problem in the family in *Coronation Street*, they get sent to Blackpool. It's still held in huge affection.' In deference to its pre-eminent role in twentieth-century popular culture, the BBC's *Strictly Come Dancing* is still filmed in the Tower Ballroom. When Bob Dylan chose a venue to perform in between London and Glasgow in 2013, it was Blackpool.

'The place is beyond intellect, it works on feelings and impulse, and they override everything else. The heart always wins in Blackpool,' says artist Tom Ireland, who grew up in the town and is now part of a group of artists who have taken over an old pub. We are talking in his studio, which looks out over the rooftops towards the sea.

'It gives itself over to people's wants and needs. It doesn't judge you – it's very forgiving. It understands shame. People will be the best and worst versions of themselves here, sometimes even at the same time, and there is no shaming of that, no judgement. It's healthy for humans to have a place like Blackpool. You don't have to dress up. The town will never die, and this year [when Covid restricted foreign travel in 2021] it has been insane, it's been rammed. It's felt like the photographs from years gone by. It's been a hot, sticky space.

'The madness is all about distraction and display. The mill owners invested in Blackpool to ensure they could recoup the money they were paying out in wages to their workforces. They got it all back. Cultures of excess and addictive behaviours are a great way to make money.'

One of Ireland's installations encapsulates some of the paradoxes of the place. He used a quote attributed to one of the town's nineteenth-century designers, 'If it wasn't for Blackpool, there would be revolution in Lancashire,' and alongside it placed a transcript of a speech on freedom which Thatcher gave at a party conference here, and a film of her dancing in the Tower Ballroom with a young Conservative. The town served as a venue for party conferences for several decades, until it was abandoned by both the Conservatives and Labour in the early 2000s because of inadequate facilities.

'So much of this town is about the space for individual freedom,' suggests Ireland. 'But there is always a question as to what kind of freedom is being celebrated. Blackpool's history lies in the Lancashire mills. People were punished by their boss for fifty weeks of the year, and then for two weeks they came here where there were no rules at all. That had a social value – and it still does now. It allows people to be base. Blackpool is a political tool; it's smart to allow such spaces to exist.

'People come in times of crisis and shame, and they come because they remember their childhoods here. I'm acutely aware that there's a fine line between joyful abandonment and reckless self-destructiveness. The place has a high rate of substance abuse; people are always drinking at two in the afternoon, and that's fine for holidaymakers. It's a great place to be a young person – partying and having a good time – but that can quickly become problematic. A culture of excess is encouraged.'

Ireland points to the work of Tracey Emin as an example of an artist who knows well the role of shame at the seaside resort. He points to the acquisition of one of her works by Blackpool's art gallery in 2015: a sentence in red neon with the words 'I know, I know, I know.'

'It's as if visitors leave their shame here – the excess, drunkenness and lives that aren't working out.'

From my cubbyhole of a hotel bedroom, I have a view over the struggling palms in the back gardens of the neighbouring street. Most of the night, I listened to seagulls fighting over leftovers in the alley. Drinking tea early next morning, a vignette unfolded of the confusing contradictions of Blackpool life. I was in one of the poorest wards of the town, with levels of deprivation which are amongst the highest in the country, yet from a doorway opposite, a five-year-old child emerged in a smart blue uniform of blazer, white shirt, tie, pleated skirt and hat (this last detail was the private-school giveaway). Hopping with intense excitement, the child dashed back inside the modest terraced house before re-emerging with two siblings, also in uniform. Later, Ian Treasure, who grew up in the area, said it was a familiar story; when he was a child, he had a friend who was the son of a Chinese restaurant owner who worked seven days a week to pay for the private school which gave his son the best chance of getting away.

Children who do well at school tend to leave, while thousands of people move in every year, many with mental-health problems, long-term health conditions, and drug and alcohol addiction. When the former boarding houses and hotels lost their visitors, they were bought up at rock-bottom prices to be converted into bedsits as HMOs (Houses in Multiple Occupation). Blackpool's cheap housing has acted as a magnet for those falling on hard times across the north-west over the last thirty years as the region struggled with sharply uneven economic development and a contracting welfare state. The deindustrialisation of the old mill towns, and the gentrification of Manchester and its soaring property prices, drove those struggling to keep up into Blackpool; the annual influx is around 8,000, and, of that total, two thirds (5,000) are on housing benefit and nearly half (44 per cent) are single males. Across the city, the rates of transience are high, with people moving frequently between rental properties, stretching even further the health, social-services and education systems.

Blackpool's schools face a Sisyphean task; some primary schools in the poorer wards see up to a third of their pupils change every year. Paul Turner, Blackpool's assistant director of education, tells me that in September 2021 600 kids had to be allocated places at short notice after moving into the town over the course of the summer holidays. In some schools 75 per cent of the children are on free school meals, he adds, and school budgets struggle to fund the much-needed support staff for the high number of children with special needs. He complains about the unfair funding formula which affects seaside resorts and sees Blackpool awarded £4,617 per primary pupil and £6,223 per secondary pupil in 2022/23, compared to inner London boroughs such as Islington, which receives £5,717 and £7,629 respectively. Furthermore, with local authorities looking for affordable provision for children with special needs and local schools with a good reputation, Blackpool attracts eligible children from all over the country, even as far away as London, but there is no extra funding. Staff recruitment and retention is a particular challenge. Turner argues that educational attainment in the city is improving, and the size of the Pupil Referral Unit, once the biggest in the country, has fallen, but he acknowledges there is still a significant gap; at one school, the reading age at year seven can be three years behind. 'We won't solve it in the next year, we need a long run at it to turn the corner by 2030,' he acknowledges. Meanwhile, those parents determined to give their children the best chance scrape together the family resources to put them into private schools.

Blackpool has always known poverty; in 1910 it was estimated that around 15 per cent of the workforce were forced to turn to soup kitchens and special relief funds in the winter to feed their family. The 'heyday' of the seaside resort had a shadow side of low-pay and seasonal work, which created an economy of feast and famine. In the winter unemployment rates could reach 30 per cent. In the peak season, plentiful low-skill work attracted incomers, who lived in poor-quality accommodation, sometimes even taking shifts in beds to keep the rent low. Infant and maternal mortality was high in a place where women worked long hours during the boarding-house season. Robert Tressell's novel, *The Ragged Trousered Philanthropists*, portrayed the precarious livelihoods and brutal exploitation of the builders and decorators employed in the expanding resorts; his fictional Mugsborough was based on Hastings, but it would have been equally true of Blackpool. John Walton concludes that in many resorts like Blackpool, 'the majority of the inhabitants existed in a state of perpetual poverty which in many cases bordered on destitution'. What made their plight worse was that resorts lacked the community and family ties evident in inland towns, which mitigated severe poverty.

Hardship and indulgence have jostled alongside each other for decades in Blackpool, and the attempt to hide the former behind the glitter and the neon has never been entirely successful. In the 1980s the decline in visitor numbers jeopardised the small family-run businesses which formed a large part of the Blackpool economy; the mini recession of 1989/90 was the turning point, said Marsden, whose former constituency of Blackpool South was hit hard. 'Over the course of ten to fifteen years I saw how the longstanding owner-occupiers, the house-proud elderly ladies, died off, and the houses were sold and made into bedsits,' he explained. 'Small bed and breakfasts couldn't afford to update their properties, the tram dated from the 1930s, the seaside public realm – pier, Winter Garden and Tower – badly needed renovation, and the train links to neighbouring cities were even slower than when I was a boy. The Irish Sea was heavily polluted, and Blackpool's beaches were ruled unacceptable by the EU; in many cases, old-fashioned sewage systems emptying directly into the sea were to blame. People lost confidence in the town and moved away.

'The Conservative buy-to-let schemes under the then chancellor, George Osborne, encouraged the conversion of hotels into Houses in Multiple Occupation in the densely populated area of central Blackpool (the terraces were built close together to ensure easy access to the attractions). Others struggled on against all the odds. Even in 2019 there were 450 B & Bs in my former constituency, many of poor quality,' said Marsden.

The scale of Blackpool's plight is evident in the statistics: the gap in life expectancy between the most and least deprived areas is 12.3 years for men and 10.1 for women, one of the worst in the country. Thirty-eight per cent die before the age of seventy-five, the lowest life expectancy nationally, and in four wards it drops as low as 66.6 years. Disability-free life expectancy for men and women is also the lowest in the country. All the indicators of poor diet, excess weight, alcohol use and limited physical activity are significantly worse than the national average. It has the highest rates of hospital admissions for alcohol-related harm and drug use in the country, and, even before Covid, it had high rates of diagnosed severe mental illness, with hundreds of hospital admissions for intentional self-harm every year.

Over a quarter of children under sixteen live in households on absolute low income (below 60 per cent of the median income), and the average weekly wage in Blackpool is £445, a *third* below the national average. Yet employment levels are only slightly below the national average (74.6 per cent, compared to 75.9 per cent), revealing the extent of low-paid, low-skill in-work poverty. The town's graduate population is nearly half the national proportion (24 per cent, compared to 43 per cent). The number of children in care is three times the country average.

With data like this, the scale of the challenge in Blackpool is self-evident, and has been for decades. The key driver is housing, suggests Ian Treasure, who has been running a drug rehabilitation project for the last three years. To show me, he drove me round the streets of south Blackpool, pointing to the telltale signs of the HMOs: the long line of doorbells at the front door, and the rows of wheelie bins in the front gardens. 'It's possible to arrive on the train and find a flat for £60 a week – substantially cheaper than Manchester, Preston or Liverpool,' he says. Investors have snapped up these old properties for a pittance (around £90,000 for a six-bedroom house) and many never bother to visit, he explains. 'They buy them online, turn them into bedsits, hire a management company to rent them out (mostly to tenants on benefits) and see returns of over 8 per cent on their investment. Usually, they don't maintain the properties, and their tenants often don't have the life skills to look after them, and a spiral of decline follows.'

In the 2021 Chief Medical Officer's annual report on coastal towns, the case study on Blackpool acknowledged the endemic problem of its privately rented housing market, warning that 'with no discernible link between housing subsidy and the size or quality of accommodation, [it] creates perverse incentives for landlords to pack as many small units into their properties as possible. The outcome of this is that inner Blackpool now houses *the single most vulnerable population in the country in the most inappropriate accommodation*, [my italics] compounding disadvantage.'

The report went on to identify a worrying recent develop-ment, so called 'supported housing', whereby a landlord claims to meet the support needs of the more vulnerable claimants in order to profit from higher levels of housing benefit. In April 2021 the *Blackpool Gazette* reported concerns among councillors that landlords could charge up to £355 a week for the provision of 'supported housing', compared to £85 for standard accommoda-tion, and were doing little or nothing to warrant the extra money.

In December 2020 the property website Zoopla put Blackpool second only to Glasgow for buy-to-let, with yields of nearly 9 per cent, assuring prospective buyers that low property prices – sale prices were regularly below those advertised – and guaranteed rent made Blackpool an attractive investment; the *Guardian* reported that 70 per cent of agreed sales in Blackpool in November 2020 were buy-to-let, in a boom encouraged by the stamp-duty holiday designed to support the housing market during the pandemic.

The result is a grossly distorted housing market underpinned by state benefits. The cost to the taxpayer of Blackpool's dysfunctional housing market was put at £350 million a year in 2013, the only time it has been calculated, in a report for the right-wing think tank the Centre for Social Justice. That figure would be considerably higher by 2020, even before the impact of Covid and the cost-of-living crisis of 2022.

Thousands of landlords benefit from a secure income provided by the state, which is effectively funding tragedy. The town's director of public health acknowledged in a painfully stark comment that the low life expectancy is 'driven by young people dying young in the failed private rented housing of inner Blackpool'. The council has tried every intervention available under the law, he added, but nothing has been sufficient to 'reverse the fundamental dynamics that continue to drive the intensification of deprivation'.

Blackpool council has lobbied central government for more powers to intervene in its housing market, but with limited results. One project involving a collaboration of lawyers, police and housing officials to identify and force negligent absentee landlords to improve accommodation or face compulsory purchase has had some success in creating better-quality housing for young families, but it is an expensive and labour-intensive process, and a fraction of what is needed to turn round these blighted neighbourhoods.

Blackpool's cheap housing has created a further issue of concern; in recent years a large number of children's homes have opened in the city, run by private companies. Since 2016, twenty-nine have opened, bringing the total in two Blackpool constituencies to thirty-nine, with many more in adjacent areas.

Because of the national shortage of places in children's homes, some of those sent to Blackpool may be hundreds of miles away from their original homes. Both campaigners for children in care and Blackpool residents have become increasingly alarmed at the flow of vulnerable young people into a city already struggling with deep deprivation and drug abuse. Blackpool council instituted new restrictions in 2021 on planning permission for new homes to curb the spread, in a year in which five new homes opened. Sandcastle Care, one of the biggest operators in Blackpool, was founded by a local property developer and now runs fifty-three homes in the area; it was recently sold to Waterland Private Equity. Its homes get good Ofsted ratings, but Ofsted's Chief Inspector, Amanda Spielman, told the BBC of her concerns at the privatisation of children's social care: 'We have companies now in the market who really don't have very much interest in childcare, sometimes buying and selling homes or groups of homes quite rapidly, and building groups of homes in specific areas that aren't necessarily where the children who most need those places live or even close to where they live.'

Well away from the seafront, in a quiet area of Blackpool lined with the offices of housing management companies, Tracey Brabin heads up the local Citizens Advice Bureau. She is one of those attempting to cope with the fallout from Blackpool's severe deprivation. Depression and poor mental health are the most common issues occurring in their caseload, she says, usually linked to debt or social isolation.

'A lot of people have very low confidence and no local family connections. It affects both the young and the elderly. They're not moving enough, and they make poor lifestyle choices around drinking, smoking and diet. They could be nineteen-year-olds who do not set foot outside their bedroom, and that affects their mental and physical health, or people in their sixties with an alcohol problem, eating low-cost, poor-quality food. Healthy eating is very difficult in this town, with a lot of cheap takeaways and discount stores; you need a car to get to the supermarkets on the outskirts of town and often people don't have one.' Brabin admits it is hard to stay optimistic, and the question she keeps asking herself is, why has this been happening for so long?

'We've got to do something. During Covid, we've seen such an increase of social problems. More food parcels are being delivered, and we try to help with budgeting and debt, but people don't have enough to live on. They can be working and claiming Universal Credit, and still they don't have enough money. The work is seasonal and poorly paid, and the benefit system doesn't have the flexibility to deal with that. It doesn't take long after losing a job to get into debt. The benefit system is not fit for purpose, and it doesn't act as a safety net, so people can't lift themselves out of poverty. If you haven't enough money to heat your home and buy food, then you're not in a state to look for work. Fuel poverty is a huge issue, because the town is freezing in winter when the cold wind is coming in from the Irish Sea. We have £50,000 to give out over the winter to help with utility arrears, but it's a drop in the ocean.'

Brabin is about to start a PhD at nearby Lancaster University on health inequality. Her determination to find answers has only been strengthened by what she has witnessed in the seven years she has worked in Blackpool. 'The local director of public health talks about "shit life syndrome", a cycle that young people can't see ways to break out of, and there is a lack of ambition and aspiration. The bright young people feel they have to leave Blackpool to do well. The polarised politics in this country mean that we have missed an opportunity to think how coastal towns could change.'

Ian Treasure agrees, adding that local services are over-whelmed by demand. The project he led on homelessness and substance abuse was funded by the National Lottery, and was filling the gaps left by the overstretched local-authority drug service. Treasure says the statutory services face an impossible task. 'They can't do anything meaningful. Most of their time is taken up with risk assessment, and clients are just grilled with questions.' Treasure's project had the funding to take a different approach and patiently build trust with the people they were trying to help.

'The first time one of our workers approaches, the client usu-ally tells them to fuck off. Next time, perhaps you have thirty seconds if you are lucky. People with multiple disadvantages have had lots of professionals asking them what's the matter; instead, we asked them, "What matters to you?" We bought one guy a fishing rod, and that was the turning point in building trust. These people have often been traumatised several times over. In some ways, it's even worse to be at the seaside, because they are surrounded by people who are enjoying themselves.

'Our work was all about kindness. You can't do that kind of outreach for more than two to three years before burning out. You are listening to constant stories of abuse, and it requires tenacity; it's always two steps forward and one step back. We had a lived experience group, who provided feedback and support if someone had a relapse.'

An evaluation of the project estimated that it saved other services such as the NHS and the police £10,000 per client, but Treasure acknowledges, 'When you talk about cost savings from an intervention, it's theoretical – you can't give the hospital less money, it just means less strain on the system.' He is proud of how the project helped people, and describes how clients vol-unteered to renovate a crazy-golf course. 'It was really about the conversations. Do something alongside people and you find out what matters to them. Do something *to* people – for example, grill them via a formal assessment – and there is automatically an imbalance of power.' Treasure found himself mixing concrete alongside volunteers weeding and repairing the course. 'One volunteer admitted after a few months that it was the longest he had been out of prison in thirty years. At the launch of the golf course in May 2021 with the mayor, he turned up wearing a suit – I didn't recognise him.'

The extent of local deprivation has mobilised a lot of Blackpool residents, Treasure adds, who work on a wide range of projects such as food banks, but the need is huge. Having worked in public health in the town over many years, one of Treasure's biggest frustrations is the short-termism of funding and policy. 'I know the statistics in Blackpool – high rates of smoking, teen pregnancy and so forth, and you need at least seven years to make any impact, but usually a strategy is set for three to four years. Back in 2005 the council, NHS and police all cooperated on a 2020 vision, but it petered out in austerity. Most of the work we do is the dustpan and brush at the bottom of the cliff; we need to get better at not letting people drop off the cliff.'

Blackpool's plight has attracted large EU grants, which have transformed major parts of its infrastructure, such as the sea-front. The Tower has been extensively renovated, and a new plaza has been created, known as the Comedy Carpet; jokes by famous comedians are laid into the big pavement. It's fun, and visitors clearly were enjoying it. Around the church in the town centre, a new cobbled square has recently been built, complete with tubs of flowers and cafes. The efforts are impressive and effective, but, only a street away, the Winter Gardens, scene of many major political speeches at party conferences, is swathed in scaffolding, a shadow of its former glory; a pawnbroker has moved into one of its street-front shops. Blackpool was built at scale making renovation challenging; the dilapidation is only ever a few streets away. New plans are underway for another round of major investment. Three hundred million from the Conservative government's Town Deal (although it does not replace the enormous cuts to local-authority budgets since 2010) will help fund yet another renovation programme, featuring the city's first five-star hotel and a Blackpool Museum of Entertainment. The aim is to widen the clientele and appeal to more middle-class families, and to shift the emphasis away from the heavy-drinking party culture. 'If the plans for renovation come off, how does that trickle down?' Brabin asks, reflecting a wider nervousness about whether gentrification would bring any benefits, given the town's entrenched deprivation.

'I love Blackpool, and the promenade is amazing. I'm really proud of the EU money which built it. On a sunny day, there is nowhere better to be,' says Treasure, who has spent most of his life on the Fylde coast. The promenade sweeps sinuously along the coast, with steps down into the churning brown water at high tide. Since most of Blackpool is less than two metres above sea level, it also provides some protection against rising sea levels, as Treasure grimly points out. He's a poet in his spare time, and offers to read aloud two of his poems as we sit in a cafe outside Blackpool's former solarium, a beautiful art deco building. They powerfully express his mix of disgust and affection for his home town.

As I walk along the promenade early on a sunny day, the red-brick façade of the newly renovated Blackpool Tower glows against the blue sky. Another early riser seems less appreciative as he staggers along: 'You English, you oughter to be ashamed of yer' sain,' he shouts angrily at the waves and the seagulls. A homeless woman sleeping in a shelter, surrounded by her bags, has just woken and is lighting a cigarette as she gazes out at the glittering sea, beaming.

A group of retired people sitting on a bench by the North Pier tell me they are visiting for the day from their caravan site up the road. They remember the place with affection from childhood holidays. 'I was one of six children, and the whole family slept in one room during our holiday here. We spent all our time on the beach, because we couldn't afford any of the attractions. We loved it,' says one woman, who grew up in Wigan. Her companion, from Preston, says his family hadn't been able to afford a Blackpool holiday or even a day trip; as one of ten, there were just too many mouths to feed, and it was not until he was a young man that he visited for the first time. They all agree emphatically that, in comparison with their memories from the 1940s and 1950s, 'the place is a bit run-down and shabby', adding that they aren't quite sure why they decided to come, and are heading back to their caravan site as soon as the bus arrives.

Another couple from Penzance have stopped off at Blackpool on a coach tour heading to the Lakes. They were curious to visit the place they had loved in the 1960s and 1970s, when they came with their children, but 'it's gone downhill', they say, adding that if it hadn't been for Covid, they would be travelling to one of their favourite holiday destinations, Disney Florida or Las Vegas. 'Poor Blackpool,' concludes another visitor, shaking her head sadly. Of my random sample of visitors strolling on the promenade and pier, only a teenage couple who have grown up in the town said they liked the place, but quickly qualified that by adding that there were too many drugs. 'The dealers will sell to anyone,' says the girl, who has just started an art course at the local college. Her friend works in a call centre, but wants to travel. I asked where. 'Yorkshire or somewhere like that,' he replied, as if the county was another country, rather than a two-hour train journey over the Pennines.

My sample was unrepresentative, but a more comprehensive survey by the tourist board, Visit Britain, exposed the same paradox: people visit Blackpool, but they don't much like what they find. Only 18 per cent of visitors in Blackpool questioned were regulars, and would be coming back, while 42 per cent were defined as 'rejectors'. Either they had never visited or would never return. Visitors liked the attractions, but on every other measure Blackpool scored badly – for service, food and drink – as well as on attributes such as stylishness, romance, beauty, relaxing, authentic, and 'out of the ordinary'. One professional working in the town admitted that, 'Many in Lancashire take the view that Blackpool has crap food, and is full of people who come to drink, and the beach is covered in litter, so they don't want to bring their children.' She added that neighbouring Lytham St Annes has much more to offer, including a Michelin-star restaurant. 'Is Blackpool what tourists really want, or is it all they can get? The town is rooted in the past.'

Despite these negatives, the visitors keep coming. Tom Ireland attributes this in large part to what he calls 'the delicious treat of nostalgia'. Blackpool's past popularity ensures it can harvest a stock of childhood memories for another couple of decades. Ireland even questions some aspects of the narrative of decline. 'If you look at the photos of Thurston Hopkins from the late 1950s and 1960s, the town looks like a shithole, and this was the so-called heyday. There was always deprivation here; it has always relied on underpaid work, and the food has always been awful. Just look at the old footage, and the litter is terrible. People are nostalgic that it was all better in the past, but was it?'

In the 1960s the Pendle Film Society filmed Blackpool, and the edited footage is on YouTube; the crowds are smartly dressed – women in their best suits, heels and hats, some men in three-piece suits, with watch chains in their pockets. Their smiles have an innocence, an open-hearted exuberance at their seaside trip; problems of drug addiction, obesity, disillusionment and boredom with the English seaside still lay in the future. The sun is shining, the beaches are packed, the bingo and fruit machines are busy, the donkey rides and Punch and Judy attract an enthusiastic audience; the trains are crowded and fleets of coaches disgorge their happy customers. The camera pans over the men in their deckchairs reading the paper, the women with their knitting; a lot of people seem to be sleeping, still a precious treat for a class defined by its labour.

In a compilation of footage a decade later, now in colour, the promenade is still crowded, the trousers are flared, the hair and sideburns long, the rides more dramatic, the eyeshadow lurid; the voiceover talks of how 'the British come to the coast regardless of the weather' and 'for a century Britons have frozen on its sands and coped with the same old maddening problems of anoraks, coats, wind and rain'. It went on to praise Blackpool's attractions as 'a bustling, sparkling palace of delights', which may be 'heady, brash, vulgar, but [is] always entertaining'. Over footage of cars on the new motorway linking the town with Preston (opened in 1975), the commentary declares that 'steady is the flow to the fleshpots' of Blackpool, a town which 'looks to the future, it couldn't be brighter'.

Reading through the comment thread on these YouTube videos offers a sample of a generation's intense nostalgia for Blackpool; some admitted they were looking to see if they could spot their younger selves in the crowds: 'Makes me feel very sad for a lost era, seems like a million years ago'; 'Went last year and it has changed ... full of dirty bedsits ... second-hand shops ... winos ... drunks ... stag and hen nights ... illegals ... down and outs … druggies. Crazy how life has changed so quick it's a shame,' wrote Bob; 'Jesus, I'm getting nostalgic watching this, how quickly time goes, I was eleven years old and used to go to Blackpool every year from when I was a small boy until I was sixteen, I remember all the sounds on the front, the music and the lights and the people, the sun AND the rain, but especially the smells, sugar and doughnuts, fish and chips, wow'; 'I love watching this. It's so sad, like someone else said on here, it's like a lost era. I gaze at the people … sometimes pausing the film to look into their faces. Wondering if they are still alive and what life subsequently brought them,' commented 'Pauly'; 'I sound old when I say this, but I wish I could go back in time to when there was more innocence in the world,' wrote 'Weirdlyoneoffs'; 'How I yearn for those days. Life was simpler, and people, despite their troubles were far happier.'

Innocence and simplicity can easily be projected on to the past – by definition the past does not know what the present does – and nostalgia is a form of emotional bias which does not stand up to question. Contrary to the comments in the YouTube thread, the 1970s were far from simple, with tense international conflict between the Cold War powers, and political battles between trade unions and government which led to power cuts and the three-day week. None the less, nostalgia is powerful, stirring us deeply with inchoate feelings of loss. Graham Greene, referring to seaside resorts such as Brighton, commented on the appeal of 'seediness', claiming that 'it seems to satisfy, temporarily, the sense of nostalgia for something lost; it seems to represent a stage further back'. In a study on nostalgia as a political force, Alastair Bonnett argues that 'a profound sense of loss' is an inescapable part of modernity, 'tied to the experience of mobility and isolation'. Pointing out that the sociologist Max Weber put melancholy at the heart of his work, Bonnett suggests that nostalgia is the outcome of the 'rationalisation' of industrial urban societies, and 'above all [the] disenchantment of the world'. The challenge is to 'preserve a remnant of humanity from this parcelling out of the soul'. The music journalist and DJ Andrew Collins, commenting on the close relationship between the seaside and several genres of music, from pop to punk, came to an insightful conclusion: 'more than any other aspect of the English landscape, these seaside towns have come to represent an intense, endlessly renewing contract with nostalgia. Their original purpose – for holidays, retirement and convalescence – has been subsumed by the heady way in which they seem to describe their past within their present.' The many contradictions accommodated at the seaside offer 'a persuasive statement about larger changes, not just in British society, but in our deeper, allegorical sense of ourselves'.

Memories of Blackpool become the material on which to project soul, seek enchantment, and express deep unease with social and economic change. Nostalgia 'takes the measure of the distance people have fallen short in their efforts to make themselves at home in a constantly changing world', writes one historian. But nostalgia is an unstable emotion, and can tip into resentment and blame quickly, bringing 'people together in the act of enjoyably lashing out at something that everyone agrees is both pitiful and reprehensible', according to Bonnett; 'as an emotion, it lacks accuracy'. He has a point; in the comment thread on the YouTube footage of Blackpool, the nostalgia developed into something more disturbing: 'Hey look, kids, this is a country that used to exist called England.' The thread spiralled into a racist lament of loss: of a country, of London as a capital city, even of European civilisation. It became clear why the films had been prefaced by an advert for independent financial advice from Nigel Farage, the former leader of the UK Independence Party, in which he urged viewers to sign up because 'you have been cheated by forces beyond your control who have taken advantage of you'. The algorithms calculating where to place adverts for maximum effect had identified those watching vintage Blackpool footage as Farage's target market. According to those commenting, Blackpool's 'glory days' had come to stand for other things Britain was perceived to have lost, such as whiteness, a sense of community, standing in the world, and national pride. Powerful emotions which find little space for political expression else-where, such as disappointment, powerlessness, betrayal and racism, drift to the coastal edge.

The contemporary nostalgia for a perfect past on the coast is in sharp contrast with the history of the seaside resort through the nineteenth and first half of the twentieth century, when they were considered the epitome of modernity and progress. Competing with each other over how modern their amenities were, seaside resorts encouraged their visitors to adopt new behaviours. From their early advocacy of novelties such as sea bathing and sitting on the beach, they graduated to roller coasters; the resort had to innovate, to continually offer new pleasures. The last great era of seaside reinvention was the interwar period, when resorts from Margate and Hastings to Blackpool and Morecambe embarked on ambitious infrastructure projects to update their appeal. Dreamland in Margate was influenced by Coney Island, while Hastings built a new swimming complex which could accommodate 5,000 spectators. There was a new appetite for sunshine, promoted by a coalition of health professionals, designers and social reformers; the suntan became fashionable, no longer condemned as a sign of low status. A new architecture of leisure emerged, full of space and light, designed to encourage health and sociability. Such was the success of this architectural style that it was applied to many of the spaces of the 1951 Festival of Britain in London. The festival organisers described how they wanted to encourage the 'happy rhythms of democratic diversity associated with seaside holidays'. The periphery had arrived at the centre of the capital, and a new promenade along the South Bank, emulating the seafront, was one of the lasting legacies of the festival.

Seaside modernism was a spectacular swansong, offering an elegant aesthetic which, after a period of neglect, is once again deeply appreciated. The style and the habits it encouraged permanently changed how the seaside was enjoyed. Sunbathing became one of the main delights, alongside a new emphasis on athleticism; sports facilities emerged as an essential requirement of every resort, including swimming pools, diving boards, bowling greens, tennis courts and golf courses. Adverts used images of svelte men and women to associate the beach with youth, beauty and physical health. Swimming costumes were designed to 'expose as much skin to the open air as possible', advocated the Women's League of Health and Beauty, while organisations such as the Sunlight League provided the alibi of therapeutic purpose for a new, unapologetic sexuality.

Modernist architecture bequeathed many masterpieces on the British coastline – from the famous De La Warr Pavilion in Bexhill-on-Sea to the buildings of Blackpool Pleasure Beach designed by Tom Purvis and Joseph Emberton – but none, I would argue, matched Morecambe's extraordinary LMS Midland Hotel, opened in 1933. Prominently positioned in the middle of the promenade, it was an ambitious statement which put the town on a par with the most glamorous resorts in Europe. Morecambe had traditionally had the reputation of the 'Naples of the North', due to its curved promenade, spectacular sunsets over the Irish

Sea and stunning view of the Lake District; it had the extravagantly opulent Winter Gardens and the Alhambra dance hall. The LMS Midland was designed to fulfil the modernist dream, with its windows oriented to catch the sunsets and mountainous vistas across the water. Its architect, Oliver Hill, proudly declared that it was the 'first really modern hotel in the country'.

The bold white building was intended to have a 'sun-drenched newness', and its white surface was treated with crushed blue glass to catch the light, while the undersides of the windows were glazed green, and electrically polished to mirror the sunsets. Influenced by the design of ocean liners, new building techniques using concrete allowed for sinuous curves; the balconies and huge glass windows opened the interior to light bounced off the nearby sea. Inside, an imposing spiral staircase dominated the foyer, lined with reliefs carved by Eric Gill of Odysseus being welcomed from the sea by Nausicaa. Hill recruited a team of celebrated designers for every detail of the décor, from bath plugs to the staff uniform; Eric Ravilious painted murals and Marion Dorn designed carpets, marble floor mosaics, crockery and light fittings. A semicircular cafe offered a view of the Lake District across the expanse of Morecambe Bay, and, closer to hand, of the Super Swimming Stadium, a vast outdoor pool. No expense was spared.

'The Midland is one of the great statements internationally about the nature of modernism and modernity as an attitude to living and lifestyle,' concludes Michael Bracewell. It brought glamour and sophistication to the English – and, more specifically, the Lancashire – seaside resort. Such was its prestige that the French fashion designer Coco Chanel came to visit, her private plane landing on the sands of Morecambe Bay. It was cited as 'an extravagant gesture of hope at a time of great Depression'. Morecambe was a place with such high self-regard in the 1930s that it spoke not just *to* the nation, but *for* it, hosting the iconic event of the Miss Great Britain beauty pageant. More prosaically, it also staged the cutting of the biggest Christmas pudding in the Empire.

This arrival of modernism was a laboratory for the future: promoting a freedom of movement, abandoning formality and convention. Women experimented in their dress and behaviour, partly inspired by cinema; when visiting the seaside, they could compete in talent contests, and join beauty parades. The ordinary visitor could become a spectacle, even acquire fame, either dancing in the ballrooms with their huge balconies of people watching, or entering a swimming competition in one of the stadia built to seat thousands of spectators, as at Morecambe and Blackpool. The seaside resort's reputation for sexual adventure also became more explicit: 'the pleasure city of Blackpool, and even its refined annexes, seethed with sex, as it had to', wrote Anthony Burgess.

This swansong may have been magnificent, but it was brief. By the late 1950s the seaside was the stage used to explore themes of national decline. In *The Entertainer*, a film released in 1960 starring Laurence Olivier and Joan Plowright, the Morecambe seafront is bustling and its beauty contest is staged in a stadium packed with thousands of spectators, but it is about the end of an era. Set in the context of the disastrous Suez War of 1956, the soldier son leaves to fight (and is killed), while his father, Archie, attempts to revive his fading music-hall career, distracted by beauty-pageant contestants half his age. 'He's never been on telly,' comments a child surveying the poster of Archie's forthcoming performance; popular entertainment was being redefined by television, and seaside resorts were struggling to catch up. The playwright John Osborne wrote the original play after spending three months in Morecambe, working in the theatre. As in *Hindle Wakes*, the main protagonist is an independent young woman, this time with a job in London and a lover, but she sets both aside to return home to hold the family together. The film portrays the end of both the music hall and British imperial power; the protagonist's boyfriend, eager to emigrate to Africa, demands, 'Can you see any reason to stay in this cosy corner of Europe?' Meanwhile, at the half-empty music hall, they play a parody of 'Rule Britannia': 'Remember we are British! We're all broke and mustn't scrap the navy.' The director, Tony Richardson, commented in an interview that 'Archie was the embodiment of a national mood. Archie was the future, the decline, the sourness, the old ashes where Britain was heading.'

Osborne was one of the first artists to project themes of decline on to the seaside resort, and many have followed suit. In the space of two decades, the British resort pivoted from being a symbol of modernity and progress to a symbol of the past and the outdated. This decline happened in the full view of millions of visitors, as well as those attempting to make a living in these

towns – a public and tangible spectacle as buildings were closed down, upkeep abandoned and the hoardings went up. Even more deadly, the English seaside tradition became the object of ridicule: 'a people renowned for reticence and phlegmatic calm emerge once a year in ridiculous clothes, besport themselves in rowdy groups and are prepared to sing dated and highly senti-mental songs at the drop of a funny hat and at the top of their tuneless voices', wrote Anthony Hern in *The Seaside Holiday* in 1967. No resort experienced this pivot from the future to the past more dramatically than Morecambe.

As car ownership increased, holidaymakers spread further afield, and their visits to Blackpool and Morecambe became shorter. The numbers of caravans more than doubled in fifteen years from the mid-1950s to the late 1960s, badly denting the demand for accommodation, which was the life blood of the resort economy. By comparison, day trippers spent little. As the price of foreign package holidays dropped, they came within reach of working-class families. Large advertising budgets and holiday television programmes introduced audiences to formerly unimaginable destinations such as Turkey and Egypt. Smaller English resorts were particularly hard hit, and in two decades (1970–90), they lost nearly half their market. Morecambe saw the value of its tourism fall from £46.6 million in 1973 to £6.5 million in 1990. The scale of the once-celebrated palaces of mass entertainment – winter gardens, ballrooms, swimming-pool stadiums – became an impossible burden as the local ratepayer tax base shrank. One by one, they were abandoned, rented out, boarded up, bulldozed or left to crumble in the damp sea air. When the American writer Paul Theroux arrived in Morecambe on his journey round the English coastline in the early 1980s, he was horrified; 'if it had been richer, all these Victorian build-ings would have been torn down. The town was too poor to be vulgar,' he writes in *The Kingdom by the Sea*, adding, 'It aston- ished me that anyone would come here for a vacation and have fun, since it seemed the sort of place that would fill even the cheeriest visitor – me, for example – with thoughts of woe. I imagined day trippers getting off the train, taking one look and bursting into tears.'

The Midland Hotel faded, its famous white exterior stained and cracked under the assault of the salt-laden gales, and in 1998 it was finally closed. It stood as a derelict monument on the seafront for ten years, but it was lucky and escaped the bulldozers; passionate advocates ensured that its architec- tural significance was recognised, and it was renovated and reopened in 2008.

Cultural historian Vanessa Toulmin was born in 1967 and grew up in Morecambe's Winter Gardens fairground, until her family bought a guesthouse in the West End, then a smart part of the town, in the 1970s. At first, business was brisk. 'I used to collect luggage from the station, cook breakfasts – I even ran the crazy-golf course. We had Bradford Week and Glasgow Fortnight – people would come back every year, stay in the same room, have the same breakfast. I would babysit for their children. I loved the seaside in winter when I got it back; the rest of the time it was hard work. At eleven, I was cooking breakfast for fifteen people. It was much the same for any kid growing up with parents running a shop or chippie. I had a great family and I loved it – it didn't seem hard at the time.

'By the late 1970s and early 1980s, the town was changing. Morecambe used to have two theatres, a fairground, the Winter Gardens, two cinemas and an outdoor pool, but within two years it had all gone. I used to pretend I was from Lancaster when I got to university, because I was ashamed of Morecambe. What happened in the 1980s to seaside towns was catastrophic. Slums were cleared in Liverpool, Manchester and the Midlands under Thatcherism, and their inhabitants were sent to Morecambe. The town became known as the single-mother capital of England; it was total social breakdown.'

Witnessing Morecambe's dramatic decline shaped Toulmin's career as an advisor on cultural regeneration to coastal towns – she jokes that her colleagues call her 'the professor of crap seaside towns'. She has worked extensively in both Blackpool and Morecambe on regeneration projects. 'I'm the custodian, guardian, advocate, protector and lobbyist for seaside towns,' she proudly declares.

When Blackpool lost its bid for a super-casino in 2007 (on which hopes for the town's revival had been pinned for several years), it prompted Toulmin to mount a show of London cabaret and queer theatre, Showzam. 'I didn't want Blackpool to go through what Morecambe had experienced. Trees were growing out of the gutters of the Winter Gardens at the time. I helped the council to buy it, and we lobbied for millions of government funding. A local reporter said the Winter Gardens was like a toilet, but I said to him, "See it through my eyes." I reminded people of what had happened there: Frank Sinatra, Winston Churchill, Margaret Thatcher – they all came. At the time, popular entertainment wasn't seen as heritage in the UK the way it has been in the US; there was a middle-class bias against commercial mass entertainment,' says Toulmin. 'I have an ongoing battle with the Arts Council over the definition of what culture is.'

Now Toulmin has become chair of a charity renovating Morecambe's Winter Gardens, once known as the Albert Hall of the North. It's a massive task. 'It had been bought by local amateurs with a passion to save it, but the question is, how do you *run* it? There are lots of grants to make buildings beautiful, but none to help you maintain a place, apart from Arts Council grants, and then you have to put on stuff that won't generate revenue. My family tease me that my approach is "don't mess with Ness", so we cleared the debts and raised £1 million to repair the roof and building – and then Covid struck. But we are back up, and my vision is of a 2,500-seat music venue; I'm talking to promoters now. It's got the domed roof for the acoustics. It could play a major role in Morecambe's regeneration.

'My family history goes back 100 years in the town, and my nephews and nieces all live there; at the moment, the youngsters have to leave town to get a good job. That's what I want to change. We've got a skills academy linked to the Winter Gardens planned.'

Toulmin lives in Sheffield, but she keeps in close touch with Morecambe's Winter Gardens by webcam. Her hope is that the Winter Gardens' fortunes will be boosted by the Eden Project North's £125 million proposal for a major visitor attraction to compare with their famous Cornwall site: a series of huge mussel-shell-inspired atriums full of plants right beside the renovated Midland Hotel. Due to open in 2024, it has been years in development, and represents the most ambitious attempt to regenerate a seaside resort anywhere in the country. Eden Project North would be a spectacular addition to the town, a continuation in the ongoing process of renewal and reinvention, nearly a century after the opening of the magnificent Midland Hotel. In January 2022 it was granted planning permission, and in July 2022 the bid for a central-government contribution of £50 million was submitted, with a decision due in the autumn. But even a project as exciting and inspiring as this will struggle to transform a town after the traumatic history described by Toulmin.

*

Morecambe was basking in an Indian-summer heatwave, and the beach was filling up with visitors. The ice-cream vans were doing brisk business, but in the backstreets away from the seafront, in the Citizens Advice Bureau (CAB), Helen Greatorex and Joanna Young were keen to describe the legacy of its steep decline. Greatorex, head of Morecambe's CAB, was born in the deprived West End of Morecambe, and saw how former hotels were converted into Houses in Multiple Occupation over the last forty years to meet the demand for cheap bedsits, as in Blackpool. She started as a CAB volunteer in 1992, and as energy prices rose in April 2022, she warned the local newspaper in a letter: 'In all my thirty years of working for CAB, I have never seen so many people who have so few options to solve the issue of living on impossibly low incomes.' In line with other badly deprived coastal resorts, Morecambe has a large gap in life expectancy between the poorest and richest parts of the town, and some of the worst deprivation indicators in the country; one in four of the population has a limiting, long-term illness or disability, putting huge pressure on the local NHS and care services.

Greatorex met Young, a former marketing consultant, at the local food bank, where they were both trustees. She persuaded Young to take a 50 per cent pay cut to work as a benefits advisor. Young grew up in Winchester, Hampshire, one of the wealthiest parts of the country, and the job has been a shocking revelation of an unfamiliar world. The two make formidable allies, driven by a passionate, determined fury.

'At least back in the 1980s, when I was bringing up my kids in the West End, there was community support, and in the 1970s there was fun to be had as a kid, and cheap ways to get around,' says Greatorex. 'There wasn't the vilification of the poor which we have now. You could get away with being poor – you didn't get teased for not having the best shoes, the latest labels. Now in some places the children don't leave the street, they don't even go to the beach; they can't buy an ice cream.'

Morecambe is separated from the more prosperous university town of Lancaster by the wide muddy estuary of the River Lune. Unusually, Young and her academic husband crossed the river and bought a house in Morecambe Bay. The opposite sides of the river could be on different planets, says Young, who began as a volunteer for the food bank at the start of the pandemic.

'I was suddenly confronted by a room full of people who hadn't eaten for several days, who were cold, and didn't have electricity or hot water. The unfairness of it really smacked me in the face. I'd read all about this, but it's different when you're sitting opposite a mother of three children who hasn't eaten for three days, and she's white and shaking – and her kids are the same age as mine – and I have to put three sugars in her coffee to help her. Fate has dealt us such different hands. There is very little she can do without superhuman effort.

'In 2020 we distributed nearly 40,000 food parcels, and set up a project to distribute uniforms. We noticed that some people were taking six pairs of boys' trousers, and realised it was because they didn't have a washing machine. Some schools have installed a laundry room, they allow parents to charge their phones, and run a breakfast club. School is the one escape, the one safe place.

'People often come to the Citizens Advice Bureau desperate for a fridge, washing machine or dryer, but the Department of Work and Pensions says a fridge is not a priority. The other day, a nineteen-year-old client who had just had a baby didn't have a washing machine, so she had to go to the food bank just to get clean clothes.

'Austerity has made things so much worse. The way people

are spoken to – it is as if they didn't matter. It would make me furious, but people here are used to it. But it shouldn't be like this.'

As in Blackpool, the town has a high number of single males with addiction problems, who are very isolated and struggle to deal with the benefit system.

'They feel that everything and everyone is against them. It's difficult to create social glue with such a transient population. Their lives are chaotic, so they can't keep up with the Universal Credit (UC) system, because they don't have mobile phones or they run out of phone credit. Under UC, everyone has to have an online account, but many have poor literacy and few digital skills. There are no dedicated phone staff at the Department of Work and Pensions.

'A lot of people on UC are unwell or have mental problems, and there is a high chance that they will get something "wrong" on their applications. I saw a client this morning who had lost her mobile phone and, without it, she can't do anything – this CAB office is the only place where she can get help. The system treats them in a punitive way. It can take a long time to get disability benefit and, in the interim, they have nothing to live on.'

Greatorex estimates that a benefits advisor needs at least three years' experience to guide someone safely through the complexities of the benefit system without losing them money. Even then, a council tax or utility bill can suddenly push someone into debt. 'Life on UC feels very insecure and dangerous. A single mother can be left with less than £50 a week for bills, food and transport. Then there are what we call the JAMs, the Just About Managing, and it takes one broken washing machine to tip them into debt.'

Greatorex identifies two forms of poverty in the town: one is entrenched generational poverty, usually linked to health problems or to part-time, insecure work, often in care homes; the second is the transient population seeking cheap accommodation, such as ex-prisoners needing to get away from their home environment. 'There's low tolerance of that transient population. They used to be given a travel warrant to go home, but they don't go home, they just become homeless and beg. Then the charities give out tents; one couple were living in a tent and the partner froze because she had wet herself. Some of these people are very vulnerable.

'The area has a five-year waiting list for social housing, which makes it difficult for single people to get a home, and under twenty-five it's impossible. The situation of care leavers makes me cry – they have no financial support and often don't feel safe, and they bounce from one service to another.'

Even in the midst of this bleak landscape, Greatorex is struck by the resilience she witnesses. 'One of our clients with severe learning disabilities has brought up two children. I admire her so much. I've known her for thirty years. Many of us couldn't cope with that. There has been so much vilification of the poor and the disabled in the media. You can't understand the psychology of poverty unless you have experienced it – that stress produces constant cortisol.'

Greatorex was a commissioner on the Morecambe Bay Poverty Truth Commission and found it revelatory. The commission ran for two years (2018–20), bringing together those with lived experience of poverty, known as testifiers, and the professionals working in the public services and charities in the town. The intention was to listen to each other's perspectives. Based on a concept from post-apartheid South Africa, the model has been used in other deprived areas around the country, and is rooted in the principle that 'stories that speak expose new levers for change'. The guiding rule is that 'nothing about us,

without us, is for us'. With this in mind, the commission has
forged powerful new relationships.

'Many [testifiers] had experienced the stigma of struggling
with poverty, and how this is portrayed in the media,' commented the concluding report. 'Many felt categorised, like they
were boxed off as not important or completely unseen.' During
the commission hearings, 'people began to feel like they are
worth something, that their voices matter, and that they are
being treated as equals in the room'.

The commission helped generate a sense of 'solidarity and
belonging', and led to concrete proposals, such as a new site for
the significant Traveller population in the area, and changes in
how the utility companies dealt with arrears. Lancaster council
adopted a programme of culture change to shift how it works,
with the aim of treating people fairly as human beings.

Listening to testifiers made another commissioner, Imogen
Tyler, a sociologist at Lancaster University, furious at the indifference to the suffering – a result of austerity and punitive
benefit systems. 'The poverty is hidden and it has no traction on
the public debate. The voices and experiences of these people
are not seen as worthwhile – they have been discarded. I'm staggered by how dishonest the national story is. The inequalities
are structural, so local people can't do much to change them,
while the politicians have lost confidence that they can win elections on narratives about how poverty can be tackled,' says Tyler.
'Food banks have been normalised in the last decade. Lancaster
University now encourages a salary-sacrifice scheme. We have
moved back to a charity model of welfare, with a distinction
between the deserving and undeserving poor.'

In Tyler's book, *Stigma: The Machinery of Inequality*, she uses
as an epigraph a quote from a testifier known as 'Stephanie':
'Every news bulletin seemed to be calling me scrounger, fraud,

cheat or scum. I began self-harming, carving the words "failure", "freak" and "waste of space" into my arms, legs and stomach.' Tyler argues that stigma has been 'deliberately designed into systems of social provision in ways that make help-seeking a desperate task'.

Defining stigma as a form of slow and unobtrusive violence, Tyler claims that over recent decades the political rhetoric aimed at reducing benefit dependency has 'created emotional economies of shame and stigma, low self-esteem and low aspiration which lead to high rates of self-harm and addictive behaviours, which become entrenched in neighbourhoods.' She suggests that stigma not only devalues people, it devalues the places where they live, and thus 'destabilises local communities and social bonds' and 'leaves profound and permanent scars'.

Young insisted I borrow her bicycle for the day and pressed the bike-lock key into my hand while she and Greatorex ran through a list of places to visit, intent that I should understand the extraordinary contradictions which make up Morecambe Bay. Despite the nature of their work, they both love the town, and find its beauty a constant lift to their spirits.

The seafront promenade stretches several miles down to the old village of Heysham, with its ancient church of St Peter's. Here, visitors strolled through the graveyard of velvet green grass sheltered by old trees overlooking the Irish Sea. Inside the church, the waves breaking on the shore below were just audible, and the sunlight streamed through the stained glass on to the unusual hogback stone, carved with serpents and a terrified man, his hands held high, believed to date from the times of the Vikings. Above the church, a path wound through the woods before coming out on the headland by the ruins of the even older chapel of St Patrick's. Beside it lay a burial site, probably pre-Christian, facing the setting sun, where coffins – small enough to be those of children – had been carved out of the solid rock. I was reminded of the comment of the geographer Yi-Fu Tuan that until the eighteenth century the coast was 'a landscape of fear', and promontories were particularly sacred places, where humans turned to explore mysteries of life and death. The Irish Sea was spread out on three sides, glittering in the hot sun. To the south lay the gigantic structures of the Heysham 1 nuclear power station, and to the north lay Morecambe. This rock was a vantage point to survey the extraordinary conjunction of activities, past and present: of power generation, of Vikings, of the placatory piety of holy places beside the Celts' 'green desert', and of Morecambe as a resort for health and pleasure and its tragic legacy of poverty and stigma.

Nearby on the grasslands south of Heysham, Pontins' largest resort once stood at Middleton Sands; photographs from the 1960s show the rose gardens, dovecot, crowded auditorium and playground, all of which have vanished. Up here amongst the low-lying heather and withered grass, there was barely a breath of wind. I was sticky with the heat.

In their insightful essay on Morecambe, Michael Bracewell and Linder comment that 'the overlapping of the three sources of energy' – the nuclear, religious and social – represents 'a thinly buried tracery of heavily charged historical events, the accumulated significance of which becomes a form of circuitry, wiring the region with a particular kind of power'.

I needed a swim. The evening before, as a magnificent red sun dipped into the Irish Sea, setting the windows of the Midland blazing with colour, I had shared a bench on the jetty with another visitor, Alison. 'The quicksands are too dangerous, and the deep muddy channels mean you could quickly and easily lose your footing,' she warned, horrified I might want to swim in Morecambe. We dropped the subject and shared our appreciation of the sunset; she visits regularly and said that half an hour on the jetty every week gave her the strength to cope with a difficult life of mental ill health and poor housing.

But in the unusual heat I couldn't obey Alison's strictures. In the centre of Morecambe, I spotted a couple of people swimming and followed them in. The beach is no more than a narrow band of imported sand – the original beach was washed away by currents twenty years ago. The tide was going out and each step took me deeper into the silky soft mud, which squeezed between my toes. The grey-brown colour of the water was unappealing, given what I had heard of pollution, but the cool was welcome.

After my swim, I joined three women chatting next to me on the beach. Two were Romanian care workers, and one had lived in Morecambe all her life and worked in catering for the NHS. She had bought a house in the West End thirty years ago. 'It was a nice house with a big garden for the kids, but I never let the kids out without me – the risk of trouble was too great. It was a mistake – I couldn't sell it, so I was stuck.' She grimaced and blamed the council for the town's decline. 'The council lets the place get run-down, and then the government gives them money. It seems to think it should turn the place into a retirement village – there are loads of jobs in care.' Her judgements were harsh, but she needed someone to blame. She said she was pinning her hopes of change on the Eden Project. Then she added that she liked my accent – as if I was a foreigner.

Next on my tour was Morecambe's famous ice-cream parlour, with its glamorous art deco interior of mirrors and polished wooden panelling, but I was too late: no customers were admitted after 3.45 p.m. Likewise the crowded cafe on the jetty. The sun was still blazing and the promenade busy, but Morecambe shuts early. Through the windows of the Midland, I could see

cake stands laden with sandwiches and cakes, but it was fully
booked, the car park full of expensive cars. I finally found a
Mediterranean-themed wine bar with mauve velvet banquettes
under copious blooms of artificial wisteria. To add to the cultural
confusion, I was reading Jeanette Winterson's *Oranges Are Not
the Only Fruit* and trying to imagine her young protagonist at the
Morecambe Guest House for the Bereaved, where the Reverend
Bone inveigled funds to pay his estranged wife maintenance
while passing off his girlfriend as his wife.

I had one last place to visit before I handed the bicycle keys
back. Parts of Morecambe may be struggling, but along one
road heading out of town, new bungalows have been built with
elaborate wrought-iron gates and railings, heavily colonnaded
porches and large paved drives. Many of these mansions have
static caravans around the back. Morecambe's Traveller community is long-standing and has made money in the scrap-metal
trade across the north-west. Their most famous son is the boxer
Tyson Fury, the holder of two world heavyweight titles. He was
born three months premature, weighing less than one pound,
but his father named him after the boxer Mike Tyson, and told
hospital staff that his son would follow in the footsteps of his
namesake. In 2015 Fury fulfilled his father's prediction, but then
plunged into addiction and depression before bouncing back in
2020. A flamboyant and outspoken public personality prone to
impromptu singing performances, and father of six children, he
calls himself the 'Gypsy King'. He married his teenage sweet-
heart, and they have confusingly named all three sons Prince,
in line with his own royal claim. A Catholic and a born-again
Christian, Fury's extreme views on the role of women, homosex-
uality, anti-Semitism and abortion, in a Twitter feed dotted with
biblical quotations, have provoked outrage. His fortune is ranked
in tens of millions, but he insists that 'going to heaven is the

most important thing a man or woman could ever do', and takes his family to church every Sunday when he is in Morecambe. Fury's spectacular life story both incorporates and defies many of Morecambe's conflicting histories: as a genteel resort, as a place that had a brief role in defining modernity, and as a town of ruins and broken lives. His larger-than-life personality brings both fame and notoriety to his home town.

Lancashire's resorts have bequeathed a form of holidaymaking defined by its appetite for excess and indulgence. It has been exported to the Mediterranean – Toulmin maintains that 'Blackpool moved lock stock and barrel to Benidorm' – and can now be satisfied even further afield, in Las Vegas and Florida. What's left behind in Blackpool and Morecambe is 'an extraordinary capacity to recollect their history' with, 'at times, unsettling intensity', claim Bracewell and Linder. Their visitors are motivated by a combination of stubborn affection, loyalty and nostalgia. Bracewell and Linder maintain that 'the current role allotted within the cultural psyche to the seaside towns of England, more than any other landscape, [is] allegorical. The allegory is of "endings". Sunsets, season's end, work's end, life's end.' This allegory 'represents the lingering fade out of an archaic way of life; the bustle, excitement of another age. Englishness becomes a refinement of itself, part caricature, part requiem.'

On our last evening in Lancashire, Mayanja and I drive the few miles down the Fylde coast to Blackpool's genteel neighbouring resorts of Lytham and St Annes. There you find carefully trimmed privet hedges, well-maintained front gardens, and Edwardian mansions facing over the dunes to the Irish Sea. Artist Tom Ireland had laughed about the mutual hostility between these towns and Blackpool. 'The narrative is that Blackpool is horrible and Lytham and St Annes are nice, but they're much the same, it's just that people in Lytham have better teeth. Blackpool says Lytham is up itself and Lytham talks about "Blackpool-type behaviour".'

Lytham's seafront was quiet, and we walked over the dunes and out on to the beach. Evening was approaching and a fiery sun was gradually slipping to the horizon. The tide was out, exposing huge swathes of muddy sand. Weaving our way around shallow pools of water and low sandbanks, we set off towards two tiny figures on the horizon, who seemed to be at the water's edge.

The large tidal fetch on the Irish Sea offers the map-maker several possible edges: a coastline marked by roads and buildings, or the stretches of yellow sand billowing out to abut the euphemistically blue Irish Sea, or the murky green-blue of marshland, through which rivers such as the Ribble, Wyre and Lune wander, uncertain of how to find open sea. The concept of dry land along this north-west coast is elusive; the Duke of Edinburgh once drove a coach and horses across Morecambe Bay. The historic seaside resort of Southport, just north of Liverpool, has almost given up on the coast altogether; its long pier passes over mud and marsh and past shopping outlets before finally reaching the sea.

As we splashed through the beach's thin skim of water, we talked of the lines coasts do or don't make, and what shape of country they create. 'The British are a people whose identity is built on being everywhere, all over the world. On the coastal edges, that identity is a felt experience. We could go anywhere, or we will go nowhere and peer towards the horizon from the safety of our island,' says Mayanja. 'On the coast you find that restlessness, that expansiveness can shift into different forms; either it's "Let's go," or "Let go." Both are forms of escape or escapism.'

Here on the edge, the sea urges abandonment of one kind or another – of yourself, of others, of a country. The figures up ahead had vanished, and we stopped, suddenly unsure of how much further we could or should go. When we looked back to land, it had thinned to a narrow strip of red-tiled roofs beyond the marram grass. As the sun dropped to the horizon, the sand darkened and the pools of water became polished steel, and, in the path of the setting sun, peach gold. Each ridge of sand bounced back a gleam of brilliant colour. We were standing now in a painting by J. M. W. Turner – he loved this coastline, and painted five seascapes at Heysham. Now the drama of the light had become as disorientating as Blackpool's neon. When might the tide turn? How fast and far would it surge? All around us was the whisper and bubble of the lugworms, their casts pimpling the surface of the sand. Mayanja held her phone out to record the unsettling life under our feet for her next podcast. She was too young to remember the tragic deaths in 2004, when twenty-one Chinese immigrants working as cockle pickers were drowned by the rising tide, lost in fog on the sands of Morecambe Bay, and this wasn't the best moment to explain. The seductive beach-scape of now darkening gold had become disturbing. Hastily we turned back towards land.

8

Epilogue

I have reached the end of my journey and I'm heading home, but I take a detour to visit New Brighton, once famous as Liverpool's closest resort, only a ferry ride across the Mersey. Once its tower rivalled that of Blackpool, but nothing remains except a grassed area and a plaque marking the gigs which took place in the early 1960s and helped launch the Beatles. The art deco amusement arcade, the Palace, has been scheduled for demolition; even the beach has been washed away by treacherous Mersey currents.

Sheets of rain slam into the side of the car. I peer out between the windscreen wipers at the view of Liverpool docks across the churning Mersey. The promenade is deserted of pedestrians, and within a few minutes of leaving the car I am drenched. Cars have turned their headlights on in the gloom, and I can hear the rumbling of thunder. An ornate Victorian shelter is under restoration behind metal barriers, but it looks as if work paused a while ago. The pale-blue façade of the Palace is crumbling, streaked with damp. Out on a narrow spit of sand beside the

Mersey, a man and his young son are building a magnificent structure out of driftwood and old netting. Set against the cranes of the docks on the other side of the estuary, it could be a ship-wreck – or a life raft for the deluge.

New Brighton's Victoria Quarter is the object of a major regen-eration project. Here I find refuge from the weather, in a vintage tea shop called 'Remember When'. As drops of rainwater roll off my jacket on to the floor, I marvel at the embroidered tablecloths and lace mats, china tea cups and saucers, a flowery teapot and strainer for leaf tea. The walls are covered in photographs of New Brighton's glory days: the packed steamers arriving at the pier to disgorge the day trippers on to the small beaches and into the funfairs; the portraits of Miss New Brightons posing in their bathing suits.

At a table opposite, the owner, Michelle, breaks off from sort-ing her accounts to tell her story. 'It was the dream of a lifetime to run a tea shop. It had been a vintage memorabilia shop, and I just bought all the stock. I've had to move masses into storage to make room for the tables.'

She has lived in New Brighton much of her life, and lists the things she has watched disappear: the outdoor pool when she was fifteen, the à la carte hotel where she began her working life, the Tower and the Tower Ballroom. But she insists that 'even if there isn't much to do, the visitors still come. Yesterday the place was rammed and there was nowhere to park. People will always come for the beach, the Mersey and the Irish Sea.'

The wall of the toilet was lined with images of old-fashioned knitting patterns for babies, the shelves were stacked with old face-cream tins, and above my head was a drying rack hung with the heavy underwear of a bygone age. I could have stumbled on to a film set. It was also strangely moving, a reminder of how objects outlive both us and the places where they were used.

The photos on the wall of the crowded steamers, piers and promenades were another form of memorial. Nostalgia is the last thing left to sell.

It's hard to make sense of the paradoxes of the English sea-side resort: our deep affection and appreciation, alongside the neglect, decline and deprivation. A place of second chances and last chances; a place for some to realise their most cherished dreams and for others to find despair. Along one promenade, the whole range of the human condition can be evident, from joy to boredom, from pleasure to grief. Forty-eight million British tourists visited the seaside every year in the years up to Covid-19, but, despite the popularity of the resorts and the ways they nourish our sense of belonging and identity, they have been left to cope with the repercussions of the country's growing inequality. The huge numbers of visitors rarely translate into a strong local economy; instead, small businesses on tight margins manage the seasonal ebb and flow by keeping pay low. Meanwhile, incomers in search of cheap housing drive a coastal pattern of deprivation in England and Wales, in sharp contrast to both America and Australia, which have seen a steady shift of wealth and population from the centre to the coast. In England the plight of most of the major nineteenth-century resorts has become progressively worse in recent years; Professor Sheela Agarwal concludes that 'deprivation has increased in severity and extent over time'. Media coverage has relentlessly described resorts' struggles with drugs, benefit dependency and ill health. One of the most frequent questions I was asked as I researched this book was: which was the worst? When I dodged the question by talking about the beauty of Scarborough and Folkestone, and the fascinating history of Ramsgate and Morecambe, the response was politely sceptical. The sociologist John Urry argues that they are places which have been victims of a culture war, perceived by the middle class as vulgar, and ridiculed or ignored. (He also points out that the 'lack of academic interest is striking'.)

The question which keeps nagging me is that voiced by Tracey Brabin in Blackpool: why has this been going on for so long? As a participant in a seminar organised by the Centre for Coastal Communities at Plymouth University put it: 'I grew up in Bridlington in the 1980s and all the issues were evident then. Why are we still talking about it?' Agarwal, one of the most prominent academics studying the subject, commented in our interview that she is 'staggered by the lack of understanding of coastal communities. In the 1970s there was a rediscovery of urban disadvantage, in the 1980s the focus shifted to inner cities after the riots, but nothing comparable has happened to provoke wider awareness of the plight of coastal towns.'

What became clear as I talked to dozens of people who have been working on the issue of seaside resorts is that the answer to Brabin's question is a complex amalgam of neglect, policy failure and the wider challenges of the country's politics and economy. There are no quick fixes or single-bullet solutions. The House of Lords 2019 report, *The Future of Seaside Towns*, represents the most comprehensive attempt yet to consider the plight of coastal resorts, reviewing hundreds of submissions, and visiting several, including Scarborough, Jaywick, Skegness and Margate. Resorts have been 'neglected for too long', it concludes, and coastal towns 'have felt isolated, unsupported and left behind . . . regrettably the British seaside has been perceived as a sort of national embarrassment'. The report logged the failure of both main political parties over the last twenty-five years to intervene successfully in reversing the decline, pointing out that twenty years ago researchers were already warning that seaside towns are 'one of the least understood of Britain's problem areas', but were ignored. The Lords inquiry found that 'many in such communities were delighted that at last someone was listening to their concerns and worries'.

'It's a really tricky policy problem,' admits Lord Jim Knight, who sat on the inquiry and, as former Labour MP for South Dorset (which includes Weymouth), has spent a large part of the last thirty years considering the issue. 'There's no one-size-fits-all template; the success of Brighton and Hove doesn't work everywhere. Properly empowering localities – in my head, that's the heart of it, but Whitehall doesn't trust local areas as competent to use power wisely.'

There's no political appetite from either main party to consider that order of radical political change. Even relatively modest policy proposals which might mitigate the challenges faced by resorts founder in the face of government indifference. The Lords' report urged 'a sustained long-term effort to address the impact of transience on coastal towns' and lamented that a previous parliamentary inquiry by the House of Commons in 2007 had made the same argument and had been ignored. The 2019 report met a similar fate. The government's first response was rejected by the Lords committee as inadequate, and the second simply listed recent funding commitments rather than engaging with the inquiry's recommendations, such as setting up a cross-departmental Whitehall committee dedicated to coastal towns, possible Coastal Action Zones, and a discounted rate of VAT on hospitality to encourage tourism (the UK is one of only three out of thirty-six countries in Europe which has not pursued this policy).

Housing is key, and Blackpool council's lobbying efforts for new powers for local authorities to regulate the number of HMOs have been rejected as interference in the market. Other resorts facing a very different housing challenge, the blight of second homes, have also struggled to make the case for local powers to regulate. In June 2022 Whitby, in North Yorkshire, held a referendum on restricting the number of second homes, and a high turnout resulted in 93 per cent in favour; the vote had no power but was aimed at putting pressure on planning authorities and publicising how the town is being strangled of life. Meanwhile, the All Party Parliamentary Group on coastal towns advocates a National Coastal Strategy, but there is little sign that anyone in a position of power is listening.

The government's Levelling Up White Paper in February 2022 referred frequently to coastal towns, but there was less evidence that it had grasped their unique challenges; they were simply lumped in with other 'left-behind' places. The Levelling Up agenda is dominated by the framework of a north–south divide, which does not fit the profile of seaside deprivation evident in all regions. Nor does the seaside fit into the political calculations being made by Conservatives wishing to hold on to former Labour constituencies, the so-called 'red wall' seats.

The chance of real change seems distant, and there are plenty of ominous signs that seaside resorts could slip further and the deprivation worsen. A study in 2019 by the Social Market Foundation showed how the economic gap between resorts and the rest of the country was widening, and that was before the pandemic. Covid-19 hit many resorts disproportionately hard, given their economic dependence on the hospitality sector, and recession is a further threat, since the periphery experiences economic slowdowns more severely and recovery is slower. Given the profile of poverty, the cost-of-living crisis triggered by high energy prices is causing acute distress, straining the network of voluntary organisations such as food banks. Other long-term indicators are worrying; there has been a devastating fall of nearly 30 per cent in the number of young people in coastal constituencies accessing higher education between 2011 and 2018. The climate crisis adds further pressure; due to the danger of erosion and flooding, fifty-four out of eighty-two coastal resorts have declared a climate emergency. A report in June 2022 warned that far more coastal communities might have to be relocated inland than had been previously thought; as climate breakdown accelerates, sea levels are likely to rise by 35 cm by 2050. That will deter investment in affected towns.

Twenty years of policy neglect after three decades of decline exposes a systemic fault in the British political system: the degree of centralisation in Whitehall. Governments have been reluctant to give local authorities the power and funding needed to commit to long-term strategic change. That has been compounded by the lasting impact of the austerity policies of David Cameron's government, which cut deep into the capacity of local authorities to initiate partnerships with the private sector as their economic-development and tourism budgets were stripped back to the bone to safeguard statutory services. Councils cite cuts in these areas of over 40 per cent. Where funding has been available – between 2012 and 2021 the Coastal Communities Fund has made 359 grants, totalling £229 million – the complaint is that it has been piecemeal, short-term and often directed towards infrastructure. The list of recent projects sounds impressive, but the fund was heavily oversubscribed and spread resources thinly among the many towns in need: £1.2 million to Hastings for a new food court; £2.5 million to Scarborough to invest in new businesses; £1.4 million to Torbay to develop social enterprise; £1.4 million for a visitor centre at Walton-on-the-Naze. The £1.6 million to renovate the centre of Bognor Regis can do little, given the scale of the challenge.

One evaluation of the considerable amount invested in the regeneration of Hastings concluded that it had done little to shift the profile of deprivation, other than prevent further deterioration. All too often, central-government funds, such as the Town Deal and the Future High Streets Fund, have pitted coastal resorts against each other to chase money which is a fraction of what was lost in austerity cuts. Meanwhile, the funding formulae for health, education and local-authority budgets all fail to recognise the particular profile of need found on the coast.

Knight points to other factors which have put seaside resorts on the back foot, such as the position of many on the edge of predominantly rural county councils. Skegness is forty-two miles from Lincoln, Scarborough is on the edge of North Yorkshire, Morecambe on the edge of Lancashire, and Bognor Regis on the edge of West Sussex; these shire counties often have a very different economic profile, with little insight into the specific challenges of their seaside towns when decisions are being made about the allocation of resources. Resorts account for around 10 per cent of the UK population, but form a minority in each local authority (apart from the unitary authorities such as Blackpool and Brighton and Hove), and all too often they fall off the priority list. Meanwhile, central government economic policy has focused on regions and well-connected cities, on growing high-skill, high-productivity businesses, and there are always easier and safer investment options than a tired resort.

Local authorities face an ongoing dilemma about the extent to which they renovate their tourist attractions, or focus on diversification and mitigate the pitfalls of the seaside economy of low-pay, seasonal work and fickle consumer fashion. Agarwal comments that 'many policy debates continue to suggest that tourism could be a panacea to solve all the ills ... typically as a major generator of employment ... yet long-term tourism development has been a poisoned chalice, because the unskilled, low-paid and seasonal nature of employment has fashioned a major societal issue of poverty and deprivation'. The Levelling Up policy agenda under Boris Johnson had an explicit focus on improving productivity, and the tourism industry will always struggle to compete by that measure, as Torquay hotelier Brett Powis pointed out.

Local authorities also have to balance the interests of residents and of visitors; the former may want a swimming pool, the latter may prefer an art gallery or late licensing hours. As the Blackpool artist Tom Ireland put it, 'I'd hate to be on the council. It's a thankless task, when you have so many different constituencies and a fundamental division between the voters who are residents and the economy of the town as a visitor attraction; the bulk of the investment goes into ensuring the town attracts visitors, but maybe they should be spending more on social services? It's expensive to run a mass-visitor attraction, and they can't get on top of the litter here. In Munich I saw how the morning after a big festival the city was spotless; crews worked through the night to clean it. Why not here?'

Seaside towns were traditionally Conservative voting, reflecting their core economy of small businesses, but, frustrated by the lack of policy engagement with their decline, many voted for Labour MPs in 1997. Yet the coastal constituencies did not have the same kind of influence and deep connections in the Labour Party as the former coal mining and manufacturing towns, and perhaps their concerns struggled to get an adequate hearing. From 2005 onwards, voters drifted back to the Tories as they looked for political representation which might bring change, and by 2019 even stalwarts of the 1997 intake, such as Blackpool's Gordon Marsden and Chris Ruane (who represented Vale of Clwyd, which includes Rhyl, North Wales), lost their seats. These old hands from Labour's time in power can list many achievements – new infrastructure such as schools, roads, promenades – but they acknowledge it wasn't enough to reverse decline.

Resorts have struggled to make the case that they have a unique set of challenges. Their peripherality has affected their ability to organise, collaborate and articulate a strong collective voice in the national conversation. As Knight points out, Weymouth has more in common with Blackpool and Scarborough than with neighbouring towns inland, yet the necessary networks and coalitions have only begun to emerge in the last few years. Why has it taken so long? The old rivalries among resorts, or a lack of people and budgets to cover the time-consuming journeys to visit and share best practice? Ian Treasure told me it took him the best part of a day to get from Blackpool to Southend by public transport. Many seaside towns are not well connected even within their own counties, let alone to other resorts. In addition, perhaps resorts have been inhibited from publicising their plight for fear of damaging their appeal to visitors. Scarborough wants to communicate a good-news story of a town on the brink of change, not that it has one of the lowest rates of pay and highest levels of personal debt in the country. Resorts always prefer to tell of their successes in a bid to boost morale, rather than their struggles with HMOs, drug dependency and high levels of ill health. Meanwhile, the media has taken up the task of vilification: the fact that most resorts are an extraordinary, contradictory mix of both success and failure is too complex a story; instead the portrayal of places like Clacton, Skegness and Weston is part poverty-porn, part class-based contempt. The outcome of all these factors is that the worsening plight of seaside towns has failed to gain traction in the public imagination in the way that the inner cities did in the 1980s.

As I travelled, the image of the bollards and street signs collected by the nineteenth-century developer George Burt and installed around Swanage came back to mind. This architectural salvage is a wider metaphor for what is driven, or drifts, to the edge; resorts absorbing the spillover of their metropolitan neighbours – London, Manchester, Birmingham, Liverpool, Nottingham. The resorts are caught in the back-wash as the urban centres undergo rapid change. While the cities hold the economic, political and cultural power, their contradictions – of wealth and poverty – are propelled to the coastal edge. The demands which are so acutely felt in the resorts – for cheap housing, for second homes, or the wish for a new start or a slower pace of life – largely originate in urban centres. The resorts' struggle with transience, deprivation and desperation counts as collateral damage for the dynamism of cities. Combatting the resulting distortion of resorts' development is well beyond the resources of well-intentioned locals, and probably beyond specific coastal policies. The resorts illustrate the wider national failures of a brutally inadequate welfare state, growing inequality, failing public services, and a crisis of affordable housing. They magnify the harshness of national policies which assume, manifestly, that some lives don't matter. In the process, the resorts themselves have become stigmatised, in a national narrative which is impatient and contemptuous of broken lives. The homeless, ex-prisoners and those fleeing domestic violence leave the gentrified inner cities and are absorbed by the resorts, where they are 'out of sight, out of mind'. Even within a relatively small resort, the extent and depth of poverty can be hidden, as Sarah Cant found in Margate. Scattered around our coastline, concentrated in pockets of a few streets, those who don't or can't conform to social norms of employment and independence are effectively ghettoised. Punitive welfare and inadequate support services only reinforce the pattern of shame, stigma and despair. The result is the shocking health inequality for those living in coastal towns. The Chief Medical Officer's annual report in 2021 pointed to the incidence of cardiovascular disease, which is highest on the edges of England; on the accompanying map, a fringe of red ran along the coastline.

The reality of shortened lives makes the liminality of the coastal edges an often painful experience of suspended or

frustrated hope; the chance of a transition to a different future is elusive or unavailable, there is only a scrabbling to survive. The coast is the end of the line, with the sea's wide horizon ahead, and the only way out is back to the difficulties you were trying to escape. It's a very different experience to that which I discovered in Brighton as a teenager in the early 1980s, where I found flux and opportunity, resourcefulness and tolerance; another bitter paradox of the English seaside resort is the vastly contrasting experiences of the liminal edge. Despair is always ugly and frightening, and it lurks on many a resort street corner, or on the promenade, or in its shelters with their shattered glass, which smell of piss and weed.

It is characteristic of how the seaside inspires and regenerates that, against these odds, a myriad of determined people are still undertaking exciting initiatives. One of the most prominent is the designer Wayne Hemingway, seen as a breath of fresh air by many beleaguered councils in seaside resorts. His straight-talking insistence that the English resort can successfully reinvent itself for a new generation injects confidence. Hemingway turned to urban design after the success of his fashion label Red or Dead, and over the last twenty years he has worked in more than a dozen seaside resorts. His commitment and passion is rooted in his background; he was born in Morecambe, where his nan had a bed-and-breakfast hotel, and he spent much of his free time in the town until his teens. He talks affectionately of the crowded promenade, the waves of visitors coming in for Glasgow Fortnight, and the sense of the town as being in the throes of a fabulous party. Hemingway Design's work has won prizes and international media coverage and transformed run-down neighbourhoods, from Lowestoft and Margate to Boscombe in Bournemouth. He argues that the key is culture.

'Everything you do should be touched by the hand of an artist or a designer. If you do that, regeneration has a much larger chance of taking hold. It's not true that culture is just for the middle classes. I remember my mum and nan going out to dance in Morecambe and how they dressed up. The seaside resort has always been all about culture and entertainment. We go to the seaside for the best things of life. Our renovation of the amusement park, Dreamland, in Margate is for the mass market – it's not the Turner Contemporary, but it's still culture. Our First Light festival in Lowestoft has classical music *and* mass-market tribute bands – we believe culture is for every-one, otherwise you end up being Southwold [which caters for a middle-class market] and somewhat one-dimensional. We can't do that, because we are working in big places, some with popu-lations of 70,000 or more.'

Hemingway's approach has always been to identify what parts of a resort can be reinterpreted. He sums it up as 'one foot in the past, one in the future, that's the way to do it. Young people are progressive, they want something different.' His team have reno-vated 1950s buildings, shabby beach huts, and old abandoned industrial warehouses, bringing vivid colours, artisan markets, community-led activities and live music.

'Some of what we say falls on deaf ears, but we are still work-ing with almost all the coastal towns who have commissioned us, and things are happening, so our approach must work. Speed of change can be frustratingly slow – sometimes it's the people, sometimes the lack of investment. Things can be done cheaply, but if a place is leaking its creative young people, then a town can get in a bit of a pickle.

'Young people don't just want Victoriana – and when we said, Don't knock down the 1950s building in Boscombe Overstrand [in Bournemouth], eyebrows were raised, but it's a matter of understanding where taste is going. Now it's the coolest part of the town, with the best cafes and beach volleyball. In the past you were more likely to find used needles there. One of my key messages is: don't be afraid to polarise – someone's good taste is bad taste to someone else. That's not a problem.

'Cleethorpes has got all the intrinsic qualities to work – hidden gems, great beaches, creative community and more creatives moving in. It's not got lots of HMOs, it's got good leadership, and it just needs to believe in itself. The really tough challenge is Blackpool. I would love to work there. It's the hardest one, and it will take some really brave and left-field thinking, because of its sheer size. The bigger you are, the harder you fall and all that.'

In Hemingway's view, it is scandalous that councils allowed former hotels to be converted to HMOs, and the legacy is toxic. 'It was obvious what was going to happen if you allow a change of use which puts toilets in single rooms. If you change an old B & B into flats which are so small you couldn't swing a cat in them, you are not going to get families renting. It was complete and utter stupidity, backed up by the personal greed of landlords who bought the properties dead cheap. It's a big political issue, and Conservative governments won't interfere in the market. A lot of people own or invest in buy-to-let, so regulating that is not a vote winner, and many people don't realise how damaging it is.'

Many of Hemingway's projects rely on brave council leadership, and he is passionate about the importance of the public sector. 'The best thing is when the council owns their own property – they owned East Point Pavilion in Lowestoft, and in Margate they issued a compulsory purchase order to get Dreamland. The proposed Eden Project in Morecambe is on council-owned land. That means the council can be brave – but it's never easy, because in local politics there is always an argument about locals wanting the money spent fixing potholes.

'To this day, finding somewhere to stay in some seaside towns – with fast Wi-Fi, vegan breakfasts, good decor – is a challenge. It should be a given. There has been a slowness to embrace modern expectations of hospitality. They needed to move so much more quickly, with councils and chambers of commerce nudging hotel owners, to encourage and inspire them. The food needed to improve – I can name resorts where we have to take the food with us because there's nowhere decent to eat. You get sneered at as "weirdos from London". There are examples of innate stuckism – places that only appeal to people stuck in the past. There's a lot of thinking that the modern world is a bad place, and change isn't good.

'I'm getting more positive. We've seen change, and even Morecambe is creeping in the right direction.' But his confidence is cautious, because he recognises that the challenge is at a national, structural scale. 'In every town we are working, the gap between rich and poor is growing. That's a society-wide issue in the UK; our inequality is beaten only by the US and Portugal in the developed world. Any political party which seriously tries to tackle inequality in a big way doesn't seem to cut it with enough of the electorate – the ignorance of people of my age makes me want to cry.'

Getting Hemingway on board brings passion, welcome publicity and successfully regenerated neighbourhoods. But turning a whole resort around requires a long-term commitment from multiple players – central and local government, and the private sector – to tackle housing, improve infrastructure such as good Wi-Fi and train links, as well as investment in new businesses. What often drops off this regeneration menu is the task of tackling poor health and raising educational attainment. Agarwal points out that much of the government funding is 'typically public-sector, oriented towards individual, one-off projects that are often infrastructure-based, such as restoring a Victorian or Edwardian pier, rather than a package of integrated measures that are people-focused, addressing the causes and impact of multiple deprivation'. Research shows that 'multiple deprivation remains a major barrier to achieving regeneration in the most affected resorts', she notes. Creating an Italian-style piazza in central Blackpool with flowers and cafes doesn't hide the shabby poverty and homelessness only a couple of streets away. If headlines announce that Rhyl is the worst crime hotspot in Wales, or that Jaywick is the poorest place in the UK, and Blackpool not far behind, visitors won't come. Investment in *social infrastructure* is vital, so that the social fabric broken over a generation can be rebuilt.

Chris Ruane urged me to visit Rhyl, his home town, in North Wales, to understand what that investment might look like. The town once welcomed thousands of Liverpudlian holidaymakers, and its decline has had devastating consequences; it has some of the poorest wards in Wales, with a reputation for the worst crime levels in the country. Ruane has lived in the town all his life. At ten years old, he was 'casing' every Saturday morning over the summer, helping carry suitcases from the station to boarding houses and caravan sites, and then handing a proportion of his earnings over to his mother, a cleaner. He went on to become a Labour MP for the town, and later discovered that Cherie Blair came for her holidays as a child, and the father of John Prescott, former deputy leader of the Labour Party, had had a caravan there. It's hard to imagine government figures having such relationships with the town now, a reflection of both the decline of resorts and how politics has come to be dominated by the professional middle class. After nineteen years as a popular and dedicated MP (he lost his seat in 2015, was re-elected in 2017 but lost again in 2019), Ruane admits he is still disappointed at losing twice to Conservatives, given the investment in the town under Labour. He is also confused about why the town turned Brexit after being the beneficiary of large EU grants. The funding has bequeathed new schools, housing, sea defences, but it didn't alter the profile of deprivation of a transient population attracted by cheap bedsits in HMOs.

Rhyl does not have the picturesque appeal of its neighbour

Llandudno, and has struggled to reinvent itself, but Ruane is still hopeful; he gives me a tour of the town, pointing out improvements with pride, and citing future plans such as a Zip World attraction and a food court. They would be good for the town's morale, he says, but recognises that it's not clear how the benefits would reach the most deprived areas. What's needed here is patient, long-term investment in social infrastructure – people and relationships which can reverse decades of stigma and social breakdown. To illustrate Ruane's point, the last stop of his tour is a project set up by Nicky on an industrial estate adjacent to some of the most deprived wards. Nicky radiates energy and determination as he shows us round the workshop, polytunnel and Portakabin. The workshop is immaculate, with carpentry tools neatly hung on the walls and tidy workbenches, and a large flat-screen TV on the wall. It is used for YouTube videos to learn carpentry skills. Equally important is the log burner in the corner.

'That fire opens up everyone: that's the centre of the project. Sometimes we cook sausages and baps, and then we talk – the sort of talk about who's bullying who into drugs. The kids start with the hoodies over their heads, monosyllabic, but slowly they talk and then we can help.'

Nicky started with a small grant for a project on arson reduction with young children from three streets. 'I didn't want to preach at them, and we set up a Harry Potter inspired Quidditch match with hockey sticks. Then we moved to the woods, near the lake where I used to play as a child. I put in a rope swing and started building dens. It managed to unite the kids of the area, so they were playing with each other rather than fighting; the kids were very territorial, and their parents had been complicit. The Dens Project was about getting them back to how we played as kids, and away from TVs and screens. From there, we got them into carpentry and growing plants.

'I'm high-functioning ADHD and autistic, and I've had mental-health issues, so I know what it's like from experience. This project is for anyone, from seven to sixty plus, who has mental-health issues such as anxiety and panic attacks; we've had kids with challenging behaviour, as well as army veterans with PTSD.'

The council have now offered the project a plot of twenty acres of partly derelict wasteland, and Nicky showed me the plans for 'edible pathways' of fruit trees and bushes, with Welsh heritage apples such as the variety known as the Rhyl apple. There will be space for wild camping as well as vegetable growing, a pizza oven, firepit and play space.

'Some come to do their Community Payback here rather than going to prison. I want the locals to get the jobs on this project – they should be the ones running it. The older men have so many skills, and they can teach the younger kids; they might start with making a bird box. They learn at their own pace; we don't push them, there's no judgement. Often, they start the carpentry very impatient, and they just want the thing immediately. Sometimes they give up for a few weeks, then they come back.'

We sit down on recycled sofas in a Portakabin surrounded by Nicky's woodwork – a table and a reconditioned cupboard. He has another plan to upcycle furniture for sale. 'Some people who come here are criminals, but they come because they want to get away and start again. We have some really badly behaved kids who never went to school, whose parents didn't care. It starts with the poor kids not having what other kids have – trainers, Xbox – and that can lead to drug selling, but they change here. Since Covid, the kids have become even more unruly. Something has gone since we were kids – they don't have a moral compass. They have no respect for each other, no please, no thank you, and a lot of swearing – but it's just language. We've had six kids go on to further-education college, and without this project they might have ended up in jail. Now the first kids we had are coming back with their children.'

This is the slow, risky but crucial part of regeneration often overlooked in struggling seaside towns, and left to small-scale voluntary projects with a long-term commitment to the development of relationships, driven by a passion born from personal and local experience. Nicky is not a professional community worker who commutes from elsewhere. Projects like his exist in many seaside resorts; Hastings has a vibrant network of community organisations, for example. They exist on small grants, often lurching precariously from one funding round to the next.

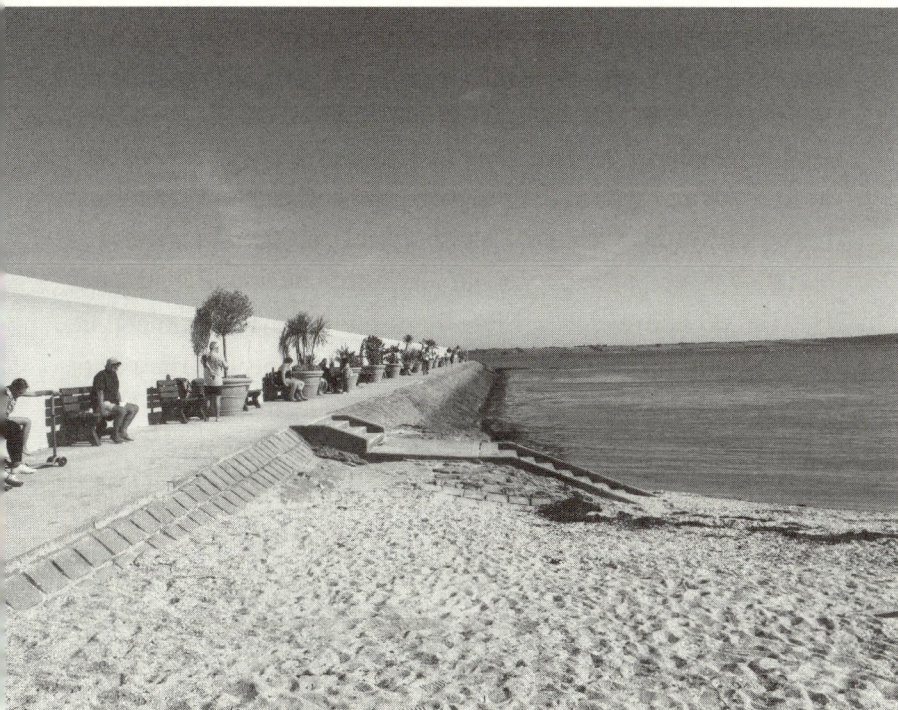

On a bigger scale, the East Quay project in Watchet, West Somerset, described in Chapter 6, shares much with Nicky's project: initiated by residents, it has strong community engagement and long-term commitment. Reflecting on eight years of working in the town, one of the founders of East Quay, Jess Prendergast, has developed a set of ideas based on their approach, which she calls Attachment Economics. At its core are three forms of attachment – to place, to neighbours, and 'through time'; this last 'reflects ancestry and shared histories of the places we inhabit, but also, crucially, our commitment to those who follow on after us. Although they appear backward-looking, efforts to protect special features in places (buildings, landscapes, culture) are an act of saving, not for the past but for the future. It represents a care being taken for those that are not yet born or have not yet moved in.'

These attachments are essential to a community's well-being and resilience, suggests Prendergast: harness them and they unleash motivation and commitment. It is through strong community connections that people feel rooted, secure and have a sense of belonging. Historically, these flourished and were supported in local businesses, churches and social groups, and 'economic and community vitality were inextricably linked', but the private sector has largely disconnected from the community, and in particular from communities on the periphery. As economic power has been centralised in the hands of multinational companies, it has lost connection to a specific place; a supermarket chain has little commitment to a community beyond the customer footfall. Places become hollowed out of economic purpose, and what is lost is the sense of value, agency and purpose, of jobs which contribute to community.

'Community is often derided as parochial and not taken seriously as an economic force. It is seen as reactionary and the opposite of "progress",' writes Prendergast in an article on the subject, but she pushes back against this and argues that 'in fact community is endlessly powerful, adaptive and kind, as proved in the pandemic. Three values stand out: compassion for others, curiosity to find new ways and learn about one another, and solidarity across divisions.' She calls for investment in the practice of building and sustaining community as the heart of all regeneration. It entails recognising that people are 'complex, messy, difficult, and relationships are works of empathy and patience. It takes time and energy and funds.'

Drawing from her experience in Watchet, Prendergast offers a manifesto of optimism. 'It suffers from economic peripherality, but there is an independence, and a deep attachment to people and the place. On many national measures West Somerset is at the bottom, but in studies of happiness and well-being we rank high. We may have no money and sometimes things are shit, but we have the things needed for well-being, such as a strong community – more so than anywhere else I've lived. Social-action projects and community groups are over 150 in a town of 4,000, and the place is built on music and theatre. We can't go out to the cinema or theatre, so we do it ourselves. We rely on each other, and on our connectedness to nature, and that's what affects our security and thus well-being. Who determines what the good life is? We shouldn't be seen as left-behind. The levelling-up agenda implicitly tells people: you must become like us. It's condescending.

'There is something about being on the edge – you are always at the end of the line, and people often don't make it all the way. It influences the culture of a place and makes it a bit more inde-pendent, even radical; it has had to find solutions. Sometimes it gets ignored, left alone by the system, but the flip side is that it has a connection to something wider, it's not insular. Often it has a history of outsiders, people coming from elsewhere, it is used to people flowing in and out. Most coastal communities have a port or are near one, and that makes them a bit more curious, welcoming, and accepting.

'Peripherality is not all negative. It's also great to be on the edge. Being by the sea means you see the horizon, and it's hard not to feel a connection to place and *to know your place*. It's hard not to be attentive to nature – waves crash on to the beach, and you see the coastguard go out to help someone. It grounds

you, and makes you constantly aware of your smallness in a bigger scheme.'

In John Gillis's history, *The Human Shore*, he argues that the global shift of populations to live on coasts across Africa, the Americas and Asia makes us 'all now creatures of the edge, mentally as well as physically. Having experienced one of the greatest physical migrations in human history, we are in the midst of a cultural reorientation of vast significance.' For centuries our preoccupation has been inland, and an ancient hesitancy about the shore is evident even in our syntax – we live *on* rather than *in* the coast – but, argues Gillis 'it turns out that change has been consistently generated on edges rather than from interiors, and therefore we need to turn our landlocked histories inside out'.

On Crosby Beach, just north of Liverpool, at every high tide human beings are slowly drowned. I watch the water creep up their thighs, reach their waist and then their chest. Further out, where the glare of hazy sunshine reflects on the sea, the dark metal of the sculptures makes them into silhouettes sharply cut from the light. I watch through my binoculars as the water reaches their chin, then their nose. Their heads appear to bob as the quiet water moves around them, gathering weight for the next wave breaking on the beach, until they begin to disappear altogether. Only a slight disturbance on the water's surface indicates the drowned statues, until even that is gone. Multiple drownings up and down the beach as far as the eye could see, and I was visiting on a day when the then home secretary, Priti Patel, had suggested that migrants attempting to cross the Channel should be turned back. This muddy north-west coast-line felt haunted, and Antony Gormley's installation, *Another Place*, gave form to the ghosts.

The beach is dotted with walkers, and, from a distance, it is hard to tell what is flesh and blood, and what is cast iron and hollow. Both seem animate. The iron ghosts are far apart – more so than photographs suggest – lonely, mute beings, cut off from each other. All stare out to sea as if in hope – of escape or rescue – towards the huge freight ships heading to the Port of Liverpool, its cranes just visible against the sky to the south. 'Stena Line connecting with Europe sustainably' runs a slogan on the cliff-like side of one ship. Their speed contrasts with the immobility of the figures.

Following Brexit, shipping between Liverpool and Dublin has steeply declined, while it has increased dramatically on direct routes from the Continent to Ireland, bypassing Britain altogether. The Merseyside coastline edges a sea between neighbours which has borne a weighty traffic of migrants fleeing hunger and looking for new opportunities. Gormley's figures stare towards the millions of ghosts who crowded the passenger ferries, coming to build industrial Britain's roads, canals and tunnels – bricklayers, coal miners, and their exhausted wives and hungry children.

Close up, the statues are rusting, and the metal suppurates in boils, flaking as if infected by some hideous skin disease. Seagulls perch on the drowning heads. I spend two hours with binoculars, watching them disappear into the muddy water, haunted by the idea that many more such drownings could be caused by sea-level rises, and that seeking asylum now exacts this cost. Coastlines as the sites of mass murder. Gormley has inverted the Romantic archetypes of solitude, the sea and the sublime, and transformed them into a statement of twenty-first-century fear of climate catastrophe, migrants and loneliness; a disturbing interpretation of the elegiac sense of endings in English seaside resorts.

Of course, if I had come a few hours later, at high tide, I would have seen a magical reverse of this gloomy tale: beings emerging from the water, erect and tall, a statement of hope, with the promise of vitality, energy and determination. A reminder that the seaside is a place of contradiction: the promise of a bigger world as well as a reminder of a small island, of beginnings as well as endings.

A year later, it's a bitterly cold February and I want a sea swim. Clevedon on the Bristol Channel is the nearest coast to where I now live, near Stroud in Gloucestershire. When I arrive, the sky is a heavy pewter grey, and a brisk wind worries the surface of the town's tidal pool; I'm a little intimidated and head off along the coastal path to warm up. Further down the coast, towards Weston-super-Mare, the Sand Point peninsula juts into the Bristol Channel and, over to the north, the lights of Cardiff are already glowing in the grey winter light. Once past the ancient church and pumping station, I stride out over the open heathland, along the edge of mud banks scarified by the water of the retreating tide. My walk reminds me of many of my previous research trips – Essex's mud, Heysham's church, Lincolnshire's caravan sites. This coast is both new to me and deeply familiar. Shafts of sunlight manage to break through, catching the edge of the waves and the contours of the glistening mud, and giving them a pale yellow sheen. Heartened, I turn back to swim – it helps if the sun is shining. One other swimmer is already in. I gasp with the shock of the freezing water; the sun has disappeared again and my swim is no more than a few breathless minutes. Bundled back into my coat, still damp, I'm chilled to the marrow, and curse this crazy habit. I need hot tea urgently. As I race along the seafront in search of a cafe, the clouds unexpectedly retreat, revealing the setting sun, which casts a glowing pink light; Clevedon's elegant pier is transformed into a strip of gleaming gold stalking over a maroon sea. Its delicate wrought-iron Chinese pavilion is silhouetted against the rose-coloured sky. Thrilled, I fall in love all over again with the seaside.

Notes

Prologue

p. 4   'In winter when the sea winds chill ...' John Betjeman, *Collected Poems*, 2006, John Murray

p. 7   'The legal position is a lot less clear than is commonly believed ...' https://www.brecher.co.uk/news/troubled-waters-rights-of-access-to-the-foreshore/

p. 8   'the Labour politician David Miliband plucked at this national heart string ...' https://www.theguardian.com/uk/2007/apr/09/greenpolitics.ruralaffairs

p. 8   'Blackpool had eighteen million visitors a year ...' https://www.visitblackpool.com/latest-news/another-year-of-growth-for-blackpool's-tourism-eco/

p. 9   'Today, 36 per cent of the UK's population ...' https://ec.europa.eu/eurostat/statistics-explained/index.php?title=Archive:Coastal_regions_-_population_statistics

p. 9   'a vision of themselves as an heroic nation ...' John Gillis, *The Human Shore*, 2012, University of Chicago Press

p. 9   'National pride reinforces the emblematic status of the seaside ...' Ursula Kluwick and Virginia Richter, *The Beach in Anglophone Literatures and Cultures: Reading Littoral Space*, 2015, Routledge

p. 10   'An English child ... cannot have long attained the power to read ...' Henry Newbolt, *Sea-Life in English Literature from the Fourteenth to the Nineteenth Century*, 1930, Nelson

p. 14   *'new research assessed that as many as 200,000 homes ...'*
           https://www.theguardian.com/environment/2022/jun/15/
           sea-level-rise-in-england-will-force-200000-to-abandon-
           homes-data-shows

p. 16   *'Liminality has a quality of ambiguity and uncertainty ...'*
           Arnold Van Gennep, *Rites of Passage*, 1961, University of
           Chicago (first published 1908)

p. 16   *'sometimes dramatic tying together of thought and experience ...'*
           Bjørn Thomassen commenting on the work of Victor Turner,
           'The Uses and Meanings of Liminality', *International
           Political Anthropology*, 2 (1): 5–27

p. 16   *'The Franciscan priest Richard Rohr ...'* https://cac.org/daily-
           meditations/liminal-space-2016-07-07/

1. Scarborough

p. 23   *'most dramatically in the US where the coastal population ...'*
           https://ec.europa.eu/eurostat/statistics-explained/index.
           php?title=Archive:Coastal_regions_-_population_statistics

p. 27   *'It was named by Historic England in 2017 ...'* https://
           historicengland.org.uk/campaigns/100-places/

p. 27   *'the history of health scares amongst guests ...'* https://www.bbc.
           co.uk/insideout/yorkslincs/series7/dirty_hotel.shtml; https://
           www.bbc.co.uk/news/uk-england-york-north-yorkshire-
           13680517; https://www.theguardian.com/uk/2002/may/10/
           health.healthandwellbeing

p. 28   *'allegations of abuse emerged against Savile's local friend,
           Peter Jaconelli ...'* https://www.thescarboroughnews.co.uk/
           news/crime/revealed-police-advised-not-take-action-
           against-ex-scarborough-mayor-jaconelli-over-1972-indecent-
           assault-1890838

p. 30   *'On the LSOA maps, tiny squares ...'* http://dclgapps.
           communities.gov.uk/imd/iod_index.html#

p. 30   *'in 2017 the town had the lowest mean employee gross salary ...'*
           Scott Corfe, *Living on the Edge: Britain's Coastal Communities*,

2017, Social Market Foundation, https://www.smf.co.uk/
publications/living-edge-britains-coastal-communities/

p. 30   *'Scarborough had the second-highest rate of personal*
insolvencies ...' https://ifamagazine.com/article/coastal-towns-
continue-to-be-hardest-hit-by-bankruptcies-seven-out-of-
the-uks-top-10-areas-for-personal-insolvency-are-by-the-sea/

p. 30   *'Scarborough's quaint streets translate into a tragic set of*
health statistics ...' North Yorkshire Joint Strategic Needs
Assessment 2019, https://hub.datanorthyorkshire.org/
dataset/1b3c5919-edec-4154-96a9-68a68991baad/resource/
e67cbea9-2934-4133-9840-467d41469f91/download/
scarborough-profile-2019.pdf; https://hub.datanorthyorkshire.
org/dataset/north-yorkshire-jsna-2019-scarborough-and-
ryedale-ccg-profile; https://fingertips.phe.org.uk/profile/
health-profiles/data#page/1/gid/1938132701/pat/6/par/
E12000006/ati/101/are/E07000168. (Some statistics in the
text have been rounded to the nearest whole figure for ease
of reading.)

p. 33   *'The American novelist John Cheever ...'* Cited in John Gillis,
The Human Shore, 2012, University of Chicago Press

p. 36   *'the idea of seeing the sea – of being near it ...'* Cited in
Elizabeth Gaskell, *The Life of Charlotte Brontë*, 1998, Penguin
Classics (first published 1857)

p. 37   *'the boundary between sea and land ...'* Rachel Carson,
The Sea Around Us, 2018, Oxford University Press (first
published 1950)

p. 38   *'The poet Sir Edmund Gosse was outraged ...'* Cited in John
Walton, *The English Seaside Resort: A Social History 1750–1914*,
1983, Palgrave Macmillan

p. 40   *'melt away when God ariseth in judgement ...'* Cited in Gillis,
The Human Shore

p. 40   *'Land's End had become a "national possession"...'* W. H.
Hudson, *The Land's End: A Naturalist's Impression of West
Cornwall*, 1923, J. M. Dent and Son

p. 44   *'Why am I so incredibly romantic about Cornwall ...'* Virginia
Woolf, 'A Sketch from the Past', *Moments of Being*, 2002,
Pimlico (first published 1976)

p. 49   *'I gazed at the shifting and melting castles ...'* Cited by Deborah
         Parsons, '"Remember Scarborough": The Sitwells on the
         Sands', in Lara Feigel and Alexandra Harris, eds, *Modernism
         on Sea: Art and Culture at the British Seaside*, 2009, Peter Lang

p. 49   *'emblematic of an entropic Edwardianism ...'* Deborah Parsons,
         '"Remember Scarborough": The Sitwells on the Sands', in ibid.

2. Skegness

p. 60   *'you can get up one morning and open your curtains ...'*
         https://www.theguardian.com/uk-news/2020/jan/18/
         its-a-monster-the-skipsea-homes-falling-into-the-north-sea

p. 60   *'In 2022 the head of the Environment Agency ...'* https://www.
         itv.com/news/anglia/2022-06-07/coastal-towns-and-villages-
         may-have-to-be-moved-inland-because-of-erosion

p. 60   *'the east coast is "soft and vulnerable" ...'* Robert Duck, *This
         Shrinking Land: Climate Change and Britain's Coasts*, 2011,
         Edinburgh University Press

p. 62   *'The district of East Lindsey (which includes Skegness) ...'*
         https://www.e-lindsey.gov.uk/article/5142/Caravan-Sites

p. 64   *'the greatest English poet of the sea ...'* Harold Nicolson,
         Tennyson, 1923, Constable and Co.

p. 64   *'How often, when a child I lay reclined ...'* *Tennyson's Suppressed
         Poems*, 1850, *Manchester Athænaum Album*, republished,
         altered, in *Life*, vol. I, p.161

p. 73   *'the crowd was less friendly and joyful than it should be ...'* Cited
         in Lara Feigel and Alexandra Harris, eds, *Modernism on Sea:
         Art and Culture at the British Seaside*, 2009, Peter Lang

p. 74   *'W. H. Auden offered advice on this new fashion of
         holidaymaking ...'* http://www.screenonline.org.uk/film/
         id/1337428/index.html

p. 75   *'a short documentary about a Skegness hotelier ...'* *Home Sweet
         Home?*, https://www.youtube.com/watch?v=fbujaWEHK8k

p. 80   *'Skegness was identified as second only to Cornwall's Newquay
         as the town most affected by Covid ...'* https://blogs.lse.ac.uk/
         politicsandpolicy/local-economic-impact-covid19/; https://ifs.
         org.uk/publications/geography-covid-19-crisis-england

p. 81   *'Over 80 per cent of residents of Skegness and Mablethorpe ...'*
https://www.gov.uk/government/publications/chief-medical-
officers-annual-report-2021-health-in-coastal-communities

p. 82   *'the area with the third highest level of anti-depressant
prescriptions ...'* https://www.theguardian.com/society/2017/
apr/14/antidepressants-prescribed-deprived-seaside-towns-
of-north-and-east-blackpool-sunderland-and-east-lindsey-nhs

p. 82   *'The area was awarded a £2.7 million National Lottery Fund
grant ...'* https://policyhub.lincoln.ac.uk/addressing-
social-isolation-among-older-people-lessons-from-talk-eat-
drink-ted-east-lindsey

p. 83   *'Life expectancy drops dramatically ...'* Lincolnshire Joint
Strategic Needs Assessment, https://www.research-lincs.org.
uk/Joint-Strategic-Needs-Assessment.aspx; https://www.gov.
uk/government/publications/chief-medical-officers-annual-
report-2021-health-in-coastal-communities

p. 84   *'despite coastal communities having an older and more deprived
population ...'* https://www.gov.uk/government/publications/
chief-medical-officers-annual-report-2021-health-in-coastal-
communities

p. 85   *'Over half the normal high tides are above the level of the land ...'*
Environment Agency, *Saltfleet to Gibraltar Point Strategy*,
https://consult.environment-agency.gov.uk/++preview++/
lincolnshire-and-northamptonshire/sgp/user_uploads/non-
technical-summary-sgp.pdf

p. 85   *'On the online national Flood Risk Map ...'* https://www.gov.
uk/check-long-term-flood-risk

p. 85   *'Sand is pumped out of the seabed and brought onshore ...'*
Environment Agency, *Saltfleet to Gibraltar Point Strategy*,
https://consult.environment-agency.gov.uk/lincolnshire-and-
northamptonshire/sgp

p. 87   *'the tidal range of the Channel shrank dramatically ...'* Nicholas
Crane, *The Making of the British Landscape*, 2016, Weidenfeld
& Nicolson

p. 89   *'The Crown Estate owns most of the UK coastline ...'* https://
www.theguardian.com/uk-news/2022/jun/16/queen-seabed-
rights-swell-value-5bn-auction-windfarm-plots-crown-estate

p. 89   '*Research seems to indicate that the kittiwakes are disorientated* …'
https://www.birdlife.org/news/2021/09/29/how-offshore-wind-
development-impacts-seabirds-in-the-north-sea-and-baltic-sea/

3. Dovercourt to Canvey Island

p. 94   '*if landscape and national identity are uneasy familiars* …'
Ken Worpole, *The New English Landscape*, 2013, Field
Station, London

p. 95   '*a ballroom with a "satin-smooth oak dance floor"* …' https://
www.harwich-history.co.uk/dovercourt-seaside-resort/

p. 96   '*town of hurry and business, not much of gaiety* …' Daniel
Defoe, *A Tour Through the Whole Island of Great Britain*, 1978,
Penguin Classics (first published 1724)

p. 97   '*Mayanja had embarked on a project analogous to my own* …'
https://carabooprojects.podbean.com/e/edging-home/; also
see her solo show at The Tetley, Leeds, in 2023

p. 111   '*the revolution started in Clacton* …' https://
www.gazette-news.co.uk/news/18972787.
douglas-carswell-brexit-revolution-started-clacton/

p. 111   '*only in Asmara after Eritrea's bloody
war* …' https://www.thetimes.co.uk/article/
tories-should-turn-their-backs-on-clacton-j0k5h6zld08

p. 111   '*he became head of a small think tank* …' https://www.
independent.co.uk/news/uk/politics/brexit-leader-to-
head-mississippi-public-policy-center-douglas-carswell-
mississippi-nigel-farage-jackson-vote-leave-b1791402.html

p. 114   '*The small resort of Jaywick started life as this kind of plotland* …'
The history is recounted in David McKie, *McKie's Gazetteer: A
Local History of Britain*, 2008, Atlantic Books

p. 114   '*I just long to see a start made on this job* …' Cited in Colin
Ward and Dennis Hardy, 'The Plotlanders', *Oral History*, Vol.
13. No. 2, *City Space and Order* (Autumn, 1985) pp. 57–70,
https://www.jstor.org/stable/40178869

p. 116   '*Perhaps one of the most bizarre chapters of Jaywick's recent
history* …' https://www.theguardian.com/us-news/2018/oct/31/
essex-street-appears-in-republican-candidate-pro-trump-advert

p. 117   *'the most deprived place in England ...'* https://www.gov.uk/
government/statistics/english-indices-of-deprivation-2019

p. 117   *'Clacton has the second-highest mental-health need ...'* https://
www.gov.uk/government/publications/chief-medical-officers-
annual-report-2021-health-in-coastal-communities; also
Joint Strategic Needs Assessment, https://data.essex.gov.uk/
dataset/exwyd/essex-jsna-and-district-profile-reports-2019

p. 118   *'given the right supports and time to evolve ...'* Cited in Stefan
Szczelkun, *The Conspiracy of Good Taste: William Morris, Cecil
Sharp and Clough Williams-Ellis and the Repression of Working
Class Culture in the C20th*, 2018, Working Press

p. 118   *'the 1960s high-rise developments were a much worse failure ...'*
Cited in Colin Ward and Dennis Hardy, *Arcadia for All:
The Legacy of a Makeshift Landscape*, 2004, Five Leaves
Publications (first published 1984)

p. 121   *'it has been a running joke for as long as I can remember ...'*
https://www.spectator.co.uk/article/southend-on-sea

p. 130   *'He declared independence and named his principality Sealand ...'*
For a riveting account of this history, see Rachel Lichtenstein's
remarkable *Estuary: Out from London to the Sea*, 2016, Penguin

p. 131   *'Oil and gas are a "submarine industry" ...'* Cited in James
Marriott and Terry Macalister, *Crude Britannia: How Big Oil
Shaped a Country's Past and Future*, 2021, Pluto Press

4. Margate to Folkestone

p. 137   *'a regular service from London Bridge, which was soon carrying
40,000 passengers a year ...'* Lee Jackson, *Palaces of Pleasure: From
Music Hall to the Seaside to Football*, 2019, Yale University Press

p. 137   *'While the inlands of England have been so hackneyed ...'* Cited
in Christine Riding and Richard Johns, *Turner & the Sea*,
2013, Thames and Hudson

p. 137   *'took on the form of a hymn to the insularity of a nation ...'* Alain
Corbin, *The Lure of the Sea: The Discovery of the Seaside 1750–
1840*, 1994, Penguin

p. 137   *'as much a place for imaginative projection as actual
visitation ...'* Matthew Ingleby and Matthew P. M. Kerr,

eds, *Coastal Cultures of the Long Nineteenth Century*, 2018,
Edinburgh University Press

p. 138 *'boundless, endless and sublime ...' Childe Harold's Pilgrimage*,
Canto 4, stanza 183, independently published 2020 (first
published 1812–18)

p. 138 *'one French critic described it as "active nothingness" ...'* Cited
in Christina Riding and Richard Johns, *Turner & the Sea*,
2013, Thames and Hudson

p. 138 *'the ocean [was] at once their nation's proper domain ...'* Samuel
Baker, *Written on Water: British Romanticism and the Maritime
Empire of Culture*, 2009, University of Virginia

p. 139 *'Of all his contemporaries [he] was the landscape artist ...'*
Christine Riding and Richard Johns, *Turner & the Sea*, 2013,
Thames and Hudson

p. 139 *'How lightly municipal, meltingly tarr'd ...'* John Betjeman,
'Margate 1940', *New Bats in Old Belfries*, 1945, John Murray

p. 141 *'parts of the centre of the town are still amongst the
most deprived ...'* https://www.gov.uk/government/statistics/
english-indices-of-deprivation-2019

p. 141 *'The average weekly wage of £528 in the district of Thanet ...'*
https://www.ramsgatetown.org/upload/docs/28%20Apr%20-
%20FGP%20Cttee%20-%20Ramsgate%20Development%20
Plan.pdf

p. 141 *'More than a third of all children in Thanet were defined as living
in poverty ...'* This figure is after housing costs; see Financial
Hardship toolkit at www.kent.gov.uk

p. 141 *'Thanet saw an increase of a third in Universal Credit claimants
during Covid ...'* An increase of 22 per cent, March 2020–
December 2021, www.kent.gov.uk

p. 142 *'the longstanding unfairness of the funding formulas ...'* https://
commonslibrary.parliament.uk/school-funding-2021-22-find-
constituency-and-school-level-allocations/

p. 143 *'In 2021 Kent council said its services for UASC ...'* https://
kccmediahub.net/kent-leader-is-saddened-to-announce-
for-the-second-time-in-a-year-we-have-reached-the-limit-
of-safe-capacity-for-the-care-of-unaccompanied-asylum-
seeking-children-and-will-take-any-action

p. 143   *'It issued legal proceedings against the home secretary ...'* https://
www.theguardian.com/uk-news/2021/jun/11/kent-county-
council-refuses-accept-unaccompanied-child-migrants

p. 143   *'Kent is a county of extremes ...'* https://www.kpho.org.uk/
joint-strategic-needs-assessment/jsna-infographics

p. 144   *'Thanet has one of the highest rates of drug-related deaths ...'*
https://geographical.co.uk/uk/uk/item/3436-margate-revival-
through-gentrification

p. 147   *'nearly half came to Margate specifically to visit the gallery ...'*
https://socialvalueuk.org/wp-content/uploads/2016/11/COaST-
Turner-SROI-2015-2016-FINAL.pdf net additional visitor
related expenditure for 2015–16

p. 150   *'On Margate Sands ...'* T. S. Eliot, *The Waste Land*, *The
Waste Land and Other Poems*, 2006, Faber & Faber (first
published 1922)

p. 152   *'The Tuggs's at Ramsgate ...'* Charles Dickens, *Sketches by Boz*,
1995, Penguin Classics (first published 1836)

p. 153   *'one of the few places "where classes mingled ..."'* Christiana
Payne, 'A Breath of Fresh Air: Constable and the Coast',
in Ingleby and Kerr, eds, *Coastal Cultures of the Long
Nineteenth Century*

p. 153   *'The coast became a "zone of cultural interchange" ...'* Valentine
Cunningham, 'On the Beach', in Ingleby and Kerr, eds,
Coastal Cultures of the Long Nineteenth Century

p. 154   *'they calmly looked on without a blush ...'* Ibid.

p. 157   *'all that was over and done with. The ring of living beauty ...'*
Cited in Paul Theroux, *The Kingdom by the Sea*

p. 161   *'The Brexit vote here was high – 64 per cent ...'* https://www.
bbc.co.uk/news/politics/eu_referendum/results/local/t

p. 165   *'it was fined a record-breaking £90 million by the Environment
Agency ...'* https://www.gov.uk/government/news/record-90m-
fine-for-southern-water-following-ea-prosecution

p. 166   *'five Thanet beaches had to be closed, including Botany Bay ...'*
https://theisleofthanetnews.com/2021/10/08/advice-not-
to-enter-water-at-14-thanet-beaches-to-remain-in-place-for-
the-weekend/

p. 166   *'In October 2021 Conservatives voted down an amendment ...'*

https://www.theguardian.com/environment/2021/oct/25/
sewage-vote-outcry-prompts-tory-mps-to-defend-decision-
on-social-media

p. 171   *'When we get out of the camp, I see people looking at us ...'*
https://www.kentlive.news/news/kent-news/inside-napier-
barracks-refugee-lifts-4618126

p. 171   *'In June 2021 the High Court ruled that the barracks ...'*
https://www.theguardian.com/uk-news/2021/jun/03/
napier-barracks-asylum-seekers-win-legal-challenge-against-
government; https://www.kentlive.news/news/kent-news/
napier-barracks-deeply-unsuitable-folkestone-7283572

p. 175   *'The sea is calm tonight ...'* Matthew Arnold, *Dover Beach and Other Poems*, 2000, Thrift Editions (first published 1867)

p. 177   *'another record-breaking summer saw 8,747 ...'* https://www.
theguardian.com/uk-news/2022/aug/28/channel-crossings-to-
the-uk-top-25000-so-far-this-year

p. 177   *'Seventeen per cent of British trade passes through this port ...'* https://
www.doverport.co.uk/port/about/chief-executives-newsletter/

p. 178   *'The White Cliffs of Dover ...'* Lyrics by Nat Burton and music by Walter Kent, 1941

5. Brighton to Bognor

p. 181   *'And there it is; the open tang, the calling ...'* Anne Enright, *The Gathering*, 2007, Jonathan Cape

p. 182   *'a carnival of strange juxtapositions ...'* John Walton, *The Victorian Seaside*, https://www.bbc.co.uk/history/british/
victorians/seaside_01.shtml

p. 183   *'As for Brighton, it corresponded with no dream ...'* Arnold Bennett, *The Clayhanger Family*, 2016, Jame-Books (first published 1910–18)

p. 184   *'the white façades of housing facing out to sea ...'* John Piper, *Brighton Aquatints*, 2020, Mainstone Press (first published 1939)

p. 184   *'an essential constituent of an English identity ...'* Frances Spalding, 'In the Nautical Tradition: John Piper', in Lara Feigel and Alexandra Harris, eds, *Modernism on Sea: Art and Culture at the British Seaside*, 2009, Peter Lang

p. 184   *'civilised constraint and liberated hedonism ...'* Walton, *The Victorian Seaside*, https://www.bbc.co.uk/history/british/victorians/seaside_01.shtml

p. 184   *'Brighton's fortunes have ebbed and flowed ...'* House of Lords Select Committee on Regenerating Seaside Towns, *The Future of Seaside Towns*, April 2019

p. 186   *'an almost sacrilegious twist on Islamic religious architecture ...'* Travis Elborough, *Wish You Were Here*, Sceptre, 2010

p. 187   *'it's a problem town, full of hippies and dropouts ...'* Malcolm Bradbury, *The History Man*, 2017, Picador (first published 1975); the television series was broadcast in 1981

p. 191   *'After Brighton, anything was possible ...'* Nicholas Watt, 'Brighton Bomber says he started the peace process', https://www.theguardian.com/uk/2000/aug/28/northernireland.nicholaswatt

p. 191   *'most audacious attack on a British government ...'* David Hughes, 'Brighton Bombing', *Daily Telegraph*, 11 October 2009 https://www.telegraph.co.uk/news/politics/6300215/Brighton-bombing-Daily-Telegraph-journalist-recalls.html

p. 193   *'The houses which looked as if they had passed through an intensive bombardment ...'* Graham Greene, *Brighton Rock*, 2004, Vintage Classics (first published 1938)

p. 193   *'no city before the war, not London or Oxford ...'* Graham Greene, *Ways of Escape*, 1999, Vintage

p. 193   *'everything including the geological formation ...'* Cited in Edwina Keown, 'The Seaside Flâneuse in Elizabeth Bowen's *The Death of the Heart*' in Feigel and Harris, eds, *Modernism on Sea*

p. 196   *'boisterous and violent but photos show laughter ...'* Rob Shield, *Places on the Margin: Alternative Geographies of Modernity*, 1992, Routledge

p. 196   *'These long-haired, mentally unstable petty little hoodlums ...'* Cited in ibid.

p. 197   *'a 1980s nursery rhyme, an ode to childhood ...'* Interview in *Performing Songwriter*, https://performingsongwriter.com/articles-interviews/cover-artist-interviews/david-bowie/

p. 198   *'marriage is a dirty joke or a comic disaster ...'* George Orwell,

'The Art of Donald McGill', 1941, *Horizon*, https://www. orwell.ru/library/reviews/McGill/english/e_mcgill

p. 199   '*the Marriage Law Reform Society argued that ...*' Cited in Claire Langhamer, 'Adultery in Postwar England', *History Workshop Journal*, 2006, 62 (1) pp. 86–115

p. 199   '*The idea of sexual licentiousness by the seaside ...*' Svetlin Stratiev, 'The Margin of the Printable: Seaside Postcards and Censorship', in Feigel and Harris, eds, *Modernism on Sea*

p. 201   '*sex, drunkenness, the loo, working class snobbery ...*' George Orwell, 'The Art of Donald McGill', 1941, *Horizon*, https:// www.orwell.ru/library/reviews/McGill/english/e_mcgill

p. 202   '*A fifth of children in the city are living in poverty ...*' http://www.bhconnected.org.uk/sites/bhconnected/ files/Brighton%20%26%20Hove%20JSNA%202017%20 executive%20summary%20VFINAL%2015%2008%2017.pdf

p. 203   '*Hastings has been the object of regeneration funds ...*' Chief Medical Officer's Annual Report, 2021, *Health in Coastal Communities*, https://www.gov.uk/government/publications/ chief-medical-officers-annual-report-2021-health-in-coastal-communities

p. 206   '*By 2016 Worthing had the second oldest population ...*' https:// www.bbc.co.uk/news/uk-43316697

p. 209   '*an outdoor photography gallery – a series entitled* Beyond Land ...' Mandy Williams, https://mandywilliams.com/ beyond-land

p. 212   '*What do you see? – Posted like silent sentinels all around the town ...*' Herman Melville, *Moby Dick*, 2022, Chartwell Classics (first published 1851)

p. 219   '*I have filthy insane digs, a great bulging scrag of a woman ...*' Cited in the *Paris Review*, Issue 39, Fall 1966

p. 219   '*Pinter said that one line of the play was the most important ...*' Interview in the *New York Times*, 30 December 1988

p. 219   '*an epic hymn to the melancholy of the English seaside ...*' Michael Bracewell, 'Morecambe: The Sunset Coast', in Feigel and Harris, eds, *Modernism on Sea*

6. Torquay to Weston-super-Mare

p. 224   '*one of the places – along with the Isle of Wight – likely to be hardest hit . . .*' A. Davenport, et al., *The Geographical Impact of Covid-19 Crisis Will be Diffuse and Hard to Manage*, 2020, Institute of Fiscal Studies, www.ifs.org.uk

p. 229   '*the 20 per cent most deprived areas . . .*' Case study, Chief Medical Officer's Annual Report, 2021, *Health in Coastal Communities*

p. 229   '*the highest proportion of children and young people with a Special Educational Need . . .*' ibid.

p. 230   '*Tragically, Torbay has a high suicide rate . . .*' Devon Community Foundation https://devoncf.com/grants/communities-local-action/

p. 237   '*The villages of Braunton and Croyde saw a 22.5 per cent surge in property prices . . .*' https://www.devonlive.com/news/devon-news/north-devon-house-prices-soar-5981950

p. 240   '*it emerged that he had worked remotely from the Caribbean . . .*' https://www.theguardian.com/politics/2021/nov/09/geoffrey-cox-under-pressure-to-quit-for-working-from-caribbean

p. 241   '*the poem that he later wrote, "The Beach" . . .*' Ted Hughes, *Collected Poems*, ed. Paul Keegan, 2003, Farrar, Straus and Giroux

p. 243   '*parts of central Ilfracombe are amongst the most deprived 5 per cent . . .*' https://devoncf.com/wp-content/uploads/2021/04/Poverty-and-Deprivation.pdf; https://consult.torridge.gov.uk

p. 244   '*The town has been hit hard by the pandemic . . .*' The Centre for Towns 2020 report found that Ilfracombe and Barnstaple were amongst the 5 per cent most economically exposed to impact of Covid. [No longer available online.]

p. 245   '*Exmoor has one of the lowest population densities . . .*' Chief Medical Officer's Annual Report, 2021, *Health in Coastal Communities*

p. 246   '*the West Somerset town has the unenviable reputation . . .*' Ibid.

p. 246   '*Twenty per cent of the population have no car . . .*' http://www.somersetintelligence.org.uk/

p. 246   '*Johnson described to Maxtone Graham a holiday regime . . .*' Cited in Ysenda Maxtone Graham, *British Summer Time*

Begins: The School Summer Holidays 1930–1980, 2020,
Little Brown

p. 247   '*Britain's social mobility problem is not just one of income ...*'
https://www.theguardian.com/society/2017/nov/28/it-feels-
a-little-forgotten-west-somerset-bears-brunt-of-social-mobility-
challenge

p. 252   '*annual turnover has risen steadily for more than a decade
to reach £241 million in 2019 ...*' https://www.statista.com/
statistics/641124/butlins-annual-revenues-united-kingdom-
uk/

p. 253   '*it had been bought back by a former owner, the Harris
family ...*' https://www.business-live.co.uk/retail-consumer/
butlins-sold-co-founders-bourne-25060672

p. 253   '*While former employees pointed out online that the job suits
some ...*' https://uk.indeed.com/cmp/Butlin's/reviews

p. 257   '*The collective raised £7 million from government funds and
foundations ...*' The government's Coastal Community Fund,
the Arts Council and the Esmée Fairbairn Foundation

p. 263   '*Deprivation is amongst the worst 1 per cent in the country ...*'
https://www.thewestonmercury.co.uk/news/20528242.cal
l-greater-scrutiny-westons-rehabs/

p. 263   '*the highest number of drug and alcohol rehabilitation centres ...*'
https://www.n-somerset.gov.uk/wp-content/uploads/2015/11/
adult-drug-misuse-chapter.pdf

p. 265   '*The gloom of the British seaside at its most dilapidated ...*' www.
theguardian.com/artanddesign/jonathanjonesblog/2015/
aug/21/in-dismaland-banksy-has-created-something-truly-
depressing

7. Blackpool and Morecambe

p. 270   '*England's quintessential seaside resort and still its most
popular ...*' https://www.local.gov.uk/case-studies/blackpool-
supplying-skills-local-visitor-economy

p. 272   '*modelled on the great European ballrooms ...*' Vanessa
Toulmin, Blackpool Tower, 2011, Boco Publishing

p. 272   '*mill-girl glided cheek by jowl with the factory master's daughter ...* ' Arthur Laycock, *Warren of Manchester*, 1900, Simpkin, Hamilton, Kent & Co

p. 273 '*foreign commentators had been astonished by the violence and crudeness ...*' J. M. Golby and A. W. Purdue, *The Civilisation of the Crowds*, 1999, Sutton

p. 274 '*crucible of conflict between classes and lifestyles ...*' John Walton, *The English Seaside Resort: A Social History 1750–1914*, 1983, Palgrave Macmillan

p. 275 '*In the 1930s researchers were alarmed ...*' ibid.

p. 276 '*portrayed as a land of dream-like happiness ...*' Lara Feigel and Alexandra Harris, eds, *Modernism on Sea: Art and Culture at the British Seaside*, 2009, Peter Lang

p. 277 '*the final solution of the periodical need for an orgy ...*' James Laveer, 'Blackpool', in Yvonne Cloud, ed., *Beside the Sea*, 1938, Bodley Head

p. 277 '*the birth of an art of excess and garish overabundance ...*' Feigel and Harris, eds, *Modernism on Sea*

p. 277 '*to the places where the crowds are, where the rhythm ...*' Gary Cross, ed., *Worktowners at Blackpool: Mass Observation and Popular Leisure in the 1930s*, 1990, Routledge

p. 278 '*Alfred Gregory's photographs of Blackpool ...*' A few are available at https://matouenpeluche.typepad. com/matouenpeluche/2011/08/alfred-gregorys-1960s-blackpool.html

p. 278 '*Blackpool was still the biggest seaside resort in Europe ...*' John Urry, *The Tourist Gaze: Leisure and Travel in Contemporary Societies*, 1990, Sage

p. 278 '*Blackpool was "out-vulgarised only by Las Vegas" ...*' Cited in ibid.

p. 278 '*artist Tom Ireland, who grew up in the town ...*' http://www. tomirelandhq.org.uk; https://abingdonstudios.org.uk/artists/tom-ireland-2/

p. 281 '*the annual influx is around 8,000 ...*' House of Lords Select Committee on Regenerating Seaside Towns, *The Future of Seaside Towns*, April 2019; and Blackpool case study, Chief Medical Officer's Annual Report, 2021, *Health in Coastal Communities*

p. 281   '*the unfair funding formula which affects seaside resorts …*'
          https://commonslibrary.parliament.uk/school-funding-2021-
          22-find-constituency-and-school-level-allocations/

p. 282   '*educational attainment in the city is improving …*' Since 2016,
          Blackpool has had one of the government's Opportunity
          Area Programmes, with the aim of improving education

p. 283   '*the majority of the inhabitants existed in a state of perpetual
          poverty …*' John Walton, *The British Seaside: Holidays
          and Resorts in the Twentieth Century*, 2000, Manchester
          University Press

p. 283   '*Blackpool's beaches were ruled unacceptable by the EU …*'
          https://www.theguardian.com/environment/2016/oct/13/
          the-eus-effect-on-blackpools-beaches-before-and-after-
          pictures

p. 284   '*The scale of Blackpool's plight is evident in the statistics …*'
          Chief Medical Officer's Annual Report, 2021, *Health in
          Coastal Communities*, Chapter 2

p. 285   '*second only to Glasgow for buy-to-let …*' https://advantage.zpg.
          co.uk/insights/articles/best-rental-yields/

p. 285   '*70 per cent of agreed sales in Blackpool in November 2020 were
          buy-to-let …*' https://www.theguardian.com/business/2020/
          dec/14/buy-to-let-sales-boom-as-landlords-rush-to-benefit-
          from-stamp-duty-holiday; Zoopla puts the figure at
          30 per cent

p. 286   '*The cost to the taxpayer of Blackpool's dysfunctional housing
          market …*' Centre for Social Justice, *Turning the Tide: Social
          Justice in Five Seaside Towns*, August 2013, https://www.
          centreforsocialjustice.org.uk/wp-content/uploads/2013/08/
          Turning-the-Tide.pdf

p. 286   '*driven by young people dying young …*' For more analysis see
          https://www.blackpooljsna.org.uk/Documents/Public-Health-
          Annual-Reports/Public-Health-Annual-Report-2017.pdf

p. 286   '*a large number of children's homes have opened in the city …*'
          Sanchia Berg and Katie Inman, *Why Have so Many Children's
          Homes Opened in Blackpool?*, https://www.bbc.co.uk/news/
          uk-62479564

p. 287   '*We have companies now in the market …*' Cited in ibid.

p. 291   '*He's a poet in his spare time and offers to read …*' He writes under the pseudonym of Ian Roustear

p. 293   '*a more comprehensive survey by the tourist board, Visit Britain …*' https://www.visitbritain.org/sites/default/files/vb-corporate/Documents-Library/documents/England-documents/destination_report_-_blackpool.pdf

p. 295   '*it seems to satisfy, temporarily, the sense of nostalgia …*' Graham Greene, *A Journey Without Maps*, 2002, Vintage

p. 295   '*In a study on nostalgia as a political force …*' Alastair Bonnett, *Left in the Past: Radicalism and the Politics of Nostalgia*, 2010, Bloomsbury

p. 296   '*more than any other aspect of the English landscape …*' https://blog.scottsmenswear.com/morrisseys-every-day-is-like-sunday/

p. 296   '*takes the measure of the distance people have fallen short …*' Peter Fritzsche, cited in Bonnett, *Left in the Past*

p. 298   '*happy rhythms of democratic diversity …*' Paul Rennie, 'Postwar Promenade', in Feigel and Harris, eds, *Modernism on Sea*

p. 299   '*The Midland is one of the great statements …*' Michael Bracewell, 'Morecambe: The Sunset Coast', in ibid.

p. 300   '*the pleasure city of Blackpool, and even its refined annexes …*' Cited in Walton, *The British Seaside*

p. 301   '*Archie was the embodiment of a national mood …*' Cited in James M. Welsh and John C. Tibbett (eds), *The Cinema of Tony Richardson: Essays and Interviews*, 1999, State University of New York

p. 306   '*one in four of the population has a limiting, long-term illness or disability …*' Chief Medical Officer's Annual Report, 2021, *Health in Coastal Communities*

p. 309   '*The commission ran for two years (2018–20) …*' https://www.morecambebaypovertytruthcommission.org.uk

p. 312   '*the coast was "a landscape of fear" …*' Yi-Fu Tuan, *Landscapes of Fear*, 2013, University of Minnesota

p. 312   '*the overlapping of the three sources of energy …*' Michael Bracewell and Linder, *I Know Where I am Going*, 2003, Bookworks

8. Epilogue

p. 322   *'deprivation has increased in severity and extent over time ...'*
Sheela Agarwal, et al., 'Disadvantage in English Seaside
Resorts: A Typology of Deprived Neighbourhoods', *Tourism
Management*, Vol. 69, December 2018, pp. 440–59

p. 322   *'places which have been victims of a culture war ...'* Cited in
Nigel J. Morgan and Annette Pritchard, *Power and Politics at
the Seaside: The Development of Devon's Resorts in the Twentieth
Century*, 1999, Liverpool University Press

p. 323   *'one of the least understood of Britain's problem areas ...'*
Christina Beatty and Steve Fothergill, *The Seaside
Economy*, 2003, https://www.shu.ac.uk/centre-
regional-economic-social-research/publications/
the-seaside-economy-the-final-report-of-the-seaside-towns-
research-project

p. 325   *'the economic gap between resorts and the rest of the
country ...'* Scott Corfe, *Falling off a Cliff*, 2019, Social
Market Foundation, https://www.smf.co.uk/publications/
falling-off-cliff/

p. 325   *'a devastating fall of nearly 30 per cent in the number of
young people ...'* House of Lords Select Committee on
Regenerating Seaside Towns, *The Future of Seaside Towns*,
April 2019, https://publications.parliament.uk/pa/ld201719/
ldselect/ldseaside/320/32002.htm

p. 325   *'far more coastal communities might have to be relocated
inland ...'* https://www.theguardian.com/environment/2022/
jun/15/sea-level-rise-in-england-will-force-200000-to-
abandon-homes-data-shows

p. 326   *'One evaluation of the considerable amount ...'* Chief Medical
Officer's Annual Report, 2021, *Health in Coastal Communities*

p. 327   *'many policy debates continue to suggest that tourism could be a
panacea ...'* Agarwal, et al., 'Disadvantage in English Seaside
Resorts', *Tourism Management*, Vol. 69, December 2018,
pp. 440–59

p. 340   *'reflects ancestry and shared histories of the places we inhabit ...'*
Jess Prendergast, *Attachment Economics: Everyday Pioneers*

for the Next Economy, May 2021, https://medium.com/
onioncollective/attachment-economics-everyday-pioneers-
for-the-next-economy-d0a9ac20080

p. 340   '*Community is often derided as parochial* . . .' Ibid.

Select Bibliography

Addey, David, *A Voyage Around Great Britain: Orkney to Southend-on-Sea, In the Footsteps of William Daniell (1769–1837)*, 2002, Spellmount

Addey, David, *A Voyage Around Great Britain: Sheerness to Land's End, In the Footsteps of William Daniell (1769–1837)*, 1995, Spellmount

Austen, Jane, *Lady Susan, The Watsons, Sanditon*, 1974, Penguin Classics (first published 1817)

Bowen, Elizabeth, *The Death of the Heart*, 2012, Vintage (first published 1938)

Bracewell, Michael, and Linder, *I Know Where I am Going: A Guide to Morecambe & Heysham*, 2003, Bookworks

Brodie, Allan, and Winter, Gary, *England's Seaside Resorts*, 2007, English Heritage

Centre for Social Justice, *Turning the Tide: Social Justice in Five Seaside Towns*, August 2013, https://www.centreforsocialjustice.org.uk/wp-content/uploads/2013/08/Turning-the-Tide.pdf

Conrad, Joseph, *Mirror of the Sea*, 2020, independently published (first published 1906)

Corbin, Alain, *The Lure of the Sea: The Discovery of the Seaside 1750–1840*, 1994, Penguin

Corfe, Scott, *Living on the Edge: Britain's Coastal Communities*, 2017, Social Market Foundation, https://www.smf.co.uk/publications/living-edge-britains-coastal-communities/

Corfe, Scott, *Falling off a Cliff*, 2019, Social Market Foundation, https://www.smf.co.uk/publications/falling-off-cliff/

Chief Medical Officer's Annual Report, 2021, *Health in Coastal Communities*, https://www.gov.uk/government/publications/chief-medical-officers-annual-report-2021-health-in-coastal-communities

Crane, Nick, *Coast*, 2010, BBC Books

Cumming, Laura, *On Chapel Sands*, 2019, Vintage

Davenport, A., et al., *The Geographical Impact of Covid-19 Crisis Will be Diffuse and Hard to Manage*, 2020, Institute of Fiscal Studies, www.ifs.org.uk

Dickens, Charles, 'The Tuggs's at Ramsgate', *Sketches by Boz*, 1995, Penguin Classics (first published 1836)

Elborough, Travis, *Wish You Were Here*, 2010, Sceptre

Feigel, Lara, and Harris, Alexandra, eds, *Modernism on Sea: Art and Culture at the British Seaside*, 2009, Peter Lang

Gillis, John, *The Human Shore*, 2012, University of Chicago Press

Golby, J. M., and Purdue, A. W., *The Civilisation of the Crowd: Popular Culture in England 1750–1900*, 1999, Sutton

Greene, Graham, *Brighton Rock*, 2004, Vintage Classics (first published 1938)

House of Lords Select Committee on Regenerating Seaside Towns, *The Future of Seaside Towns*, April 2019, https://publications.parliament.uk/pa/ld201719/ldselect/ldseaside/320/32002.htm

Ingleby, Matthew, and Kerr, Matthew P. M., eds, *Coastal Cultures of the Long Nineteenth Century*, 2018, Edinburgh University Press

Ishiguro, Kazuo, *The Remains of the Day*, 2010, Faber & Faber (first published 1988)

Kluwick, Ursula, and Richter, Virginia, *The Beach in Anglophone Literatures and Cultures: Reading Littoral Space*, 2015, Routledge

Lenček, Lena, and Gideon, Bosker, *The Beach: The History of Paradise on Earth*, 1998, Martin Secker & Warburg

Lichtenstein, Rachel, *Estuary: Out from London to the Sea*, 2016, Penguin

Maclaren-Ross, Julian, *Of Love and Hunger*, 2002, Penguin (first published 1947)

Morgan, Nigel J., and Pritchard, Annette, *Power and Politics at the Seaside: The Development of Devon's Resorts in the Twentieth Century*, 1999, Liverpool University Press

Niven, Alex, *New Model Island*, 2019, Repeater Books

Office of National Statistics (ONS), *Coastal Towns in England and Wales*, 2020, https://www.ons.gov.uk/businessindustryandtrade/tourismindustry/articles/coastaltownsinenglandandwales/2020-10-06

Osborne, Roger, *The Floating Egg: Episodes in the Making of Geology*, 1988, Jonathan Cape

Piper, John, *Brighton Aquatints*, 2020, Mainstone Press (first published 1939)

Priestley, J. B., *English Journey*, 2009, Great Northern Books (first published 1934)

Pretty, Jules, *This Luminous Coast*, 2011, Full Circle Editions

Raban, Jonathan, *Coasting*, 1995, Picador

Riding, Christine, and Johns, Richard, *Turner & the Sea*, 2013, Thames and Hudson

Shaw, Gareth, and Williams, Allan, *The Rise and Fall of British Coastal Resorts: Cultural and Economic Perspectives*, 1997, Cengage Learning EMEA

Sherriff, R. C., *The Fortnight in September*, 2006, Persephone (first published 1931)

Sitwell, Osbert, *Before the Bombardment*, 1985, Oxford Paperbacks (first published 1926)

Sitwell, Sacheverell, *A Sketch of Scarborough Sands*, unpublished, https://norman.hrc.utexas.edu/fasearch/findingAid.cfm?eadid=00238

Swift, Graham, *Last Orders*, 1996, Picador

Theroux, Paul, *The Kingdom by the Sea*, 1983, Hamish Hamilton

Toulmin, Vanessa, *Blackpool Pleasure Beach*, 2011, Boco Publishing

Toulmin, Vanessa, *Blackpool Tower*, 2011, Boco Publishing

Toulmin, Vanessa, *Blackpool Illuminations: The Greatest Show on Earth*, 2012, Boco Publishing

Tressel, Robert, *The Ragged Trousered Philanthropists*, 2004, Penguin (first published 1914)

Tyler, Imogen, *Stigma: The Machinery of Inequality*, 2020, Zed Books

Urbain, Jean-Didier (translated by Catherine Porter), *At the Beach*, 2003, University of Minnesota Press

Urry, John, *The Tourist Gaze: Leisure and Travel in Contemporary Societies*, 1990, Sage

Walton, John, *The English Seaside Resort: A Social History 1750–1914*, 1983, Palgrave Macmillan

Walton, John, *The British Seaside: Holidays and Resorts in the Twentieth Century*, 2000, Manchester University Press

Worpole, Ken, and Orton, Jason, *The New English Landscape*, 2013, Field Station, London

Worpole, Ken, and Orton, Jason, *350 Miles: An Essex Journey*, 2005, Essex Regeneration and Development Agency

Useful websites

Coastal Communities Alliance, https://www.coastalcommunities.co.uk
Centre for Coastal Communities, Plymouth University, https://www.plymouth.ac.uk/research/coastal-communities
All-Party Parliamentary Group for Coastal Communities, https://www.coastalpartnershipsnetwork.org.uk/coastalcommunitiesappg

Acknowledgements

During my seaside journeys, I have met some wonderful people and this book is immeasurably enriched by the time they were prepared to set aside to talk to me and describe their experience. Their names are in the text and they are amongst those to whom this book is dedicated. Thank you for your inspiring love of place.

As ever, I am indebted to the wonderful women who support my writing. Their encouragement, wise advice and generosity is much appreciated: my agent Sarah Chalfant is always an attentive and supportive ally in the vagaries, moments of doubt and solitary occupation of writing, and my editor Bella Lacey brings an energising enthusiasm and sense of fun without ever letting it compromise her editorial rigour and ambition. Daphne Tagg, my copyeditor and now neighbour and friend, has the sharpest eye for detail, language and structure. I am deeply grateful to Granta for its commitment to good writing and high production standards; it is always a privilege to see my work published alongside that of writers I admire. The Granta team – Christine Lo, Sarah Wasley, Pru Rowlandson, Lamorna Elmer and Simon Heafield – bring a rare dedication and passion to publishing. Kate Shearman was a meticulous proofreader.

I'm deeply grateful to Mike Savage for his insightful reading and encouragement, to Luke Wintour for some superb research and kind emergency help on picture editing, and to Samra Mayanja for her company and insights. Thank you to Ellie Wintour for two lovely photos. I owe a debt to the Society of Authors for a travel grant and to Jessica Shaw, Sally Shell and Lucy Kellaway for much appreciated hospitality on the road.

On some of my journeys I was lucky enough to be accompanied by my children. They tucked into the fish and chips and cream teas with pleasure, joined me in the freezing cold sea and Matt even jumped into Whitby harbour (it was a hot day). On several occasions, my husband was up for the adventure, be that the beauty of Gibraltar Point south of Skegness, or a bowl of seafood pasta as the sun set over Deal pier, or our chilly Kent coast dips. Thank you to my wonderful family for all their interest and patience on research trips; even as we wandered in teeming rain across Exmoor in search of petrol, Matt's support didn't waver.

Permissions

All possible care has been taken to trace the rights holders and secure permission for the texts quoted in this book. If there are any omissions, credits can be added in future editions following a request in writing to the publisher. Grateful acknowledgement is made for permission to reproduce lines of poetry or prose excerpts.

Lines from 'The Way to the Sea' by W. H. Auden, http://www. screenonline.org.uk/film/id/1337428/index.html.

Lines from 'Felixstowe, or The Last of Her Order' by John Betjeman, *Collected Poems*, 2006, John Murray; lines from 'Margate, 1940' by John Betjeman, *Collected Poems*, 2006, John Murray. Reproduced with permission of Hodder & Stoughton Ltd through PLS Clear.

Excerpt from *On Chapel Sands* by Laura Cumming, 2019, Vintage, by kind permission of the author.

Excerpt from *Wish You Were Here* by Travis Elborough, 2010, Sceptre, by kind permission of the author.

Lines from *The Waste Land* by T. S. Eliot, by kind permission of Faber & Faber Ltd.

Excerpt from *The Gathering* by Anne Enright, 2007, Jonathan Cape.

Excerpts from *350 Miles, An Essex Journey* and *The New English Landscape* by Ken Worpole, by kind permission of the author.

Photographs by the author except 'Discovering the Seashore, Runswick', p. 45, and 'Brighton Pavilion', p. 180, by kind permission of Ellie Wintour, www.elliewintour.com.

Index

Numbers in *italics* refer to illustrations.